EPA/600/R-08/014
February 2008

Effects of Climate Change on Aquatic Invasive Species and Implications for Management and Research

National Center for Environmental Assessment
Office of Research and Development
U.S. Environmental Protection Agency
Washington, DC 20460

DISCLAIMER

This document has been reviewed in accordance with U.S. Environmental Protection Agency policy and approved for publication. Mention of trade names or commercial products does not constitute endorsement or recommendation for use.

ABSTRACT

Global change stressors, including climate change and variability and changes in land use, are major drivers of ecosystem alterations. Invasive species, which are non-native species that cause environmental or economic damages or human-health impacts, also contribute to ecosystem changes. The interactions between stressors and invasive species, although not well understood, may exacerbate the impacts of climate change on ecosystems, and likewise, climate change may enable further invasions. This report reviews available literature on climate-change effects on aquatic invasive species (AIS) and examines state-level AIS management activities. Data on management activities came from publicly available information, was analyzed with respect to climate-change effects, and was reviewed by managers. This report also analyzes state and regional AIS management plans to determine their capacity to incorporate information on changing conditions generally, and climate change specifically. Although there is no mandate that directs states to consider climate change in AIS management plans, state managers can consider predicted effects of climate change on prevention, control, and eradication in order to manage natural resources effectively under changing climatic conditions. Further scientific research and data collection are needed in order to equip managers with the tools and information necessary to conduct effective AIS management in the face of climate change.

Preferred citation:
U.S. Environmental Protection Agency (EPA). (2008) Effects of climate change for aquatic invasive species and implications for management and research. National Center for Environmental Assessment, Washington, DC; EPA/600/R-08/014. Available from the National Technical Information Service, Springfield, VA, and online at http://www.epa.gov/ncea.

CONTENTS

CONTENTS (continued)

LIST OF TABLES

FOREWORD

Invasive species are a major issue ecologically and economically. Invasive species threaten native species, sometimes to the point of extinction. The economic damages and losses attributed to invasive species in the United States are sizable. In response to these issues, the federal government coordinates research and other activities concerning invasive species through the National Invasive Species Council (NISC) and the Aquatic Nuisance Species Task Force (ANSTF). NISC has written and revised a national management plan that describes strategies for prevention, early detection/rapid response, control, management, coordination, education, and research for federal agencies. The ANSTF has developed guidance for regions and states to develop their own management plan and the Task Force reviews and approves these to distribute additional funding. It is these activities that address management actions on the ground that need to incorporate climate change considerations. This report is a first step in addressing this need.

Concerns about possible climate change effects are increasingly prevalent. The U.S. Environmental Protection Agency's Global Change Research Program (GCRP) in the Office of Research and Development, National Center for Environmental Assessment, examines climate change effects on aquatic ecosystems, water quality, human health, and air quality. The results of assessments completed by the GCRP are intended to increase society's understanding of such climate change effects and contribute to decision making that improves responses to environmental challenges.

This report deals with the interaction of climate change and aquatic invasive species (AIS) and consequent effects on aquatic ecosystems. AIS are a concern to a variety of EPA programs, interagency efforts (e.g. ANSTF), state agencies, and other resource managers. This report evaluates the combined effects of climate change and AIS on aquatic ecosystem structure and function and suggests potential paths forward that will increase our understanding of these effects and improve AIS management.

Dr. Michael Slimak
Associate Director of Ecology
National Center for Environmental Assessment
U.S. EPA, Office of Research and Development

PREFACE

This report was prepared by the Environmental Law Institute (ELI) and the Global Change Research Program in the National Center for Environmental Assessment (NCEA) of the Office of Research and Development at the U.S. Environmental Protection Agency (U.S. EPA). It is intended for managers and scientists working with aquatic invasive species (AIS) to provide them with information on the potential effects of climate change on AIS and strategies for adapting their management to accommodate these environmental changes and to highlight further research needs and gaps. As a part of the information gathering for this report, U.S. EPA convened two workshops with managers and scientists. The first workshop, held at the ELI offices in Washington, DC in June 2006, focused on the current state of scientific understanding of climate-change effects on AIS and on identifying research needs and gaps. The conclusions from the first workshop led to two additional activities: (1) a review of state and regional AIS management plans to identify adaptive capacity (i.e., their ability to adjust in response to climate change) and (2) a second workshop to plan a series of review papers that addresses the connections between climate change and invasive species and the resulting complexity. The results from the review of management plans are a significant part of this report and serve as a guide for how states and regional councils may begin to incorporate climate change information into their planned activities for AIS management. The papers developed as a result of the second workshop, also held at ELI in October 2006, will be published as a Special Section in the journal *Conservation Biology*, expected June 2008.

AUTHORS AND REVIEWERS

The Global Change Assessment Staff, within the National Center for Environmental Assessment (NCEA), Office of Research and Development, was responsible for the conception and preparation of this report. This document has been prepared by the Environmental Law Institute in Washington, DC, under U.S. EPA Contract No. GS-10F-0330P. Britta Bierwagen served as the Technical Project Officer, providing overall direction and technical assistance, and contributing as an author.

AUTHORS

Environmental Law Institute
Roxanne Thomas, Austin Kane, Kathryn Mengerink

U.S. EPA
Britta G. Bierwagen

REVIEWERS

This report benefited greatly from the comments and suggestions of the following reviewers.

U.S. EPA Reviewers
Joan Cabreza, Kristina McNyset, Gina Perovich, Paul Ringold, Christine Ruf

Other Reviewers
Pam Fuller (U.S. Geological Survey), Brian Helmuth (University of South Carolina), John Stachowicz (University of California, Davis)

ACKNOWLEDGMENTS

The authors thank the Global Change Assessment Staff in NCEA, particularly Susan Julius, for their input and advice. We also thank all the reviewers in state invasive species programs who corrected and augmented the information we collected for Appendix A. We very much appreciate their time and cooperation. The members of both workshops, held in June and October 2006 to discuss research needs and information gaps, also contributed greatly to the content of this report and we thank them for their time, ideas, and suggestions throughout this process.

EXECUTIVE SUMMARY

Global change stressors, including climate change and variability and land-use change, are major drivers of changes in ecosystems. Invasive species, or non-native species that cause environmental or economic damages or impacts to human health, also cause significant changes in ecosystems and to the services they provide. The effects of climate change on invasive species and their combined effects on ecosystems are not well understood, and these changes vary regionally with climate and species traits. In some instances climate change may create additional opportunities for invasion or create conditions unsuitable for certain invasive species. Consequently, the magnitude of ecological, economic, and human-health impacts of invasive species may increase, decrease, or remain the same. The level of uncertainty about specific effects of climate change is high, yet a necessary first step to address these effects is the development of management strategies that incorporate existing climate-change information and facilitate the addition of new information. In developing this report, we strove to identify the research and management intersections that can jointly address climate change and aquatic invasive species (AIS), thereby enabling effective prevention, control, and eradication under changing conditions that states could apply to ecological problems specific to their regions.

The literature review conducted for this report and summarized in the Introduction shows that important progress has been made in identifying climate-change effects on invasive species, but in order to modify resource management activities we need a more detailed understanding of effects on specific species and interactions of other stressors. An analysis of existing AIS management plans follows the Introduction. This analysis assesses the capacity of states to modify or adapt their management activities to account for climate-change effects. The assessment shows that most states currently do not explicitly consider climate change in their AIS management plans; states are not currently mandated to do so. But the assessment did show that existing mechanisms in many state plans may be used to incorporate information about how to adapt AIS management activities to potential climate-change effects. If states can adapt their management activities, they will be more likely to maximize the effectiveness and efficiency of their financial resources as environmental conditions change, while still meeting their AIS management goals. In this respect, prevention activities may be the best way to improve effectiveness and efficiency.

Finally, this report compares information needs of AIS managers with current research to determine where gaps exist. Overall, more information and research are needed on ways climate change affects:

- Most of the AIS management activities done by states,

- Each step in the invasion pathway (transportation, colonization, establishment, and spread),

- Invasive-species impacts (ecologic, economic, and human health),

- Specific invasive species and the ecosystems they invade, and

- Interacting stressors.

These topics illustrate that much more information could be incorporated into decision making. However, there are practical steps that states can take now to adapt AIS management activities to the altered environmental conditions that are projected to exist due to climate change. Initial steps are summarized into five recommendations:

(1) Incorporate climate-change considerations into leadership and coordination activities
(2) Identify new AIS threats as a result of climate change
(3) Identify ecosystem vulnerabilities and improve methods to increase ecosystem resilience
(4) Evaluate the efficacy of control mechanisms under changing conditions
(5) Manage information systems to include considerations of changing conditions

These are some of the areas where an understanding of the effects of climate change will be important to our ability to achieve stated management goals in the future.

1. INTRODUCTION

1.1. ORGANIZATION OF REPORT

This report focuses on states' research and management needs. The goal is to enable states to perform effective prevention, control, and eradication of aquatic invasive species (AIS) in a changing climate. Although numerous federal and international efforts are relevant to—and are affected by—the concepts and recommendations discussed throughout the report, this report focuses on state-level programs, plans, and activities because they play a significant role in on-the-ground invasive-species prevention and management (ELI, 2002). Furthermore, states are likely to play an important role in driving national policy on both invasive-species and climate-change issues in the years to come.

This report examines how climate change affects AIS using published information; reviews state AIS plans and activities for existing capacity to incorporate climate-change considerations into management tasks and strategies; discusses implications for resource management, including informational and data needs; and recommends further research directions based on this discussion. It is divided into four sections and five appendices. Section 1 presents the definition of global change and the United States Environmental Protection Agency's (U.S. EPA) approach to addressing global change; briefly describes current climate-change projections and the potential effects that future climate will have on ecosystems; describes the impacts that invasive species are having on the environment; and summarizes some of the existing knowledge about how climate change affects invasive-species introduction, establishment, and spread.

Many states' AIS management activities and planned action items, as they are currently structured and outlined in management plans, do not account for the projected effects of climate change as there is no legal mandate to do so. State agencies also have limited financial resources and staff time to dedicate to AIS management activities. The disconnect between invasive-species management and potential climate-change effects may undermine efforts to achieve stated ecosystem goals under changing conditions. It should be noted that, for the purposes of this report, any modifications to management activities, plans, or programs because of climate-change considerations is termed adaptation. Adapting AIS management plans and practices will allow states to better prevent and control AIS invasions under changing conditions and will additionally maximize the effectiveness and efficiency of each state dollar spent on such activities.

In Section 2 we discuss how AIS management may be affected by changes in climate and make suggestions for modifying leadership and coordination activities, prevention strategies, control efforts, and restoration to incorporate climate-change information. We give examples of several AIS that are current priorities for many states, the management practices that are used to address these species, and the role that climate change may play in the introduction, establishment, and spread of these species.

A comparison of available information in the scientific literature with the recommendations of Section 2 reveals that more scientific, multi-stressor, long-term studies are necessary to understand more fully the interaction between climate change and invasive species and more species-specific information are needed for improved resource management.

Section 3 outlines the information needs and research gaps in our understanding of the interactions between climate change and invasive species; Section 4 concludes with a discussion of management needs for research and information to manage AIS better in the context of a changing climate; the appendices focus on additional information about AIS management and an assessment of climate-change implications for AIS management plans.

1.2. GLOBAL CHANGE

Human activities have immense impact on the global environment, and these impacts will continue if current trends persist (IPCC, 2007; MEA, 2005; Vitousek et al., 1997a). Human-induced changes are currently the primary drivers of ecosystem changes (Vitousek et al., 1997b). Global drivers of ecosystem change can include both direct drivers (e.g., climate change, nutrient pollution, land conversion that changes habitats, overexploitation, and invasive species) and indirect drivers (e.g., demographic, economic, sociopolitical, scientific, technological, cultural, and religious) (Nelson, 2005). Invasive species also follow direct changes, such as invasive species that exploit recently disturbed habitats (Didham et al., 2005). Of the direct drivers, the terrestrial environment has been most affected by land conversion to other uses, often to agricultural use (Nelson, 2005). Overexploitation of fishing resources, pollution, and climate change are examples of major drivers of change in marine ecosystems (Hughes et al., 2003; Nelson, 2005). Primary drivers of change for freshwater ecosystems include modifications and use of watersheds, human contamination of water resources, altered hydrology, and invasive species (Vitousek, 1994; Nelson, 2005). Many assessments have recognized climate change as a major driver of change that will play an increasingly important role in the coming decades (IPCC, 2007).

Global change as defined by the U.S. Global Change Research Act of 1990 (GCRA), global change "means changes in the global environment (including alterations in climate, land productivity, oceans or other water resources, atmospheric chemistry, and ecological systems) that may alter the capacity of the Earth to sustain life" (Public Law, 101-606 §2[3]). In enacting this law, Congress made the following findings, among others:

- "Industrial, agricultural, and other human activities, coupled with an expanding world population, are contributing to processes of global change that may significantly alter the Earth habitat within a few human generations," and

- "Such human-induced changes, in conjunction with natural fluctuations, may lead to significant global warming and thus alter world climate patterns and increase global sea levels. Over the next century, these consequences could adversely affect world agricultural and marine production, coastal habitability, biological diversity, human health, and global economic and social well-being" (GCRA, §101[a]).

The U.S. EPA is one of several U.S. agencies and organizations that is conducting global change research. The U.S. EPA's Global Change Research Program (GCRP) in the Office of Research and Development is assessing the effects of global change on aquatic ecosystems and their services in the context of other stressors and human dimensions in order to improve society's ability to respond and adapt to the future consequences of global change. The GCRP emphasizes the role of climate change, climate variability, and land-use change as global change stressors. Increasingly, scientists and policy-makers have recognized invasive species as global stressors because of their significant effect on ecosystems (Mooney and Hobbs, 2000; Vitousek et al., 1997a).

1.3. INVASIVE SPECIES AND ECOSYSTEM IMPACTS

The movement of species into new areas is a natural phenomenon that has occurred throughout evolutionary history (Tinner and Lotter, 2001; Graham et al., 1996). In modern times the movement of species has been augmented by humans operating in a globalized world. In the Great Lakes, for example, intense vessel traffic from international trade is the major vector for introduction of non-native aquatic species. This region has the highest known introduction rate for freshwater ecosystems, with one new non-native species being discovered every 28 weeks (Ricciardi, 2006). San Francisco Bay has the highest overall introduction rate as recorded from 1961 to 1995, with one new non-native species introduced every 14 weeks (Cohen and Carlton, 1998). The actual number of non-native species introduced into the U.S. is unknown. Estimates range from 6,600 since European settlement of the U.S. (Cox, 1999) to 50,000 species (Pimentel et al., 2005).

Non-native species (also described as alien, exotic, or nonindigenous species) that are intentionally or unintentionally released into new environments can become invasive species. Executive Order 13112 (February 1999), which established the National Invasive Species Council, defines an invasive species as a non-native species the introduction of which causes or will likely cause harm to the economy, environment, or human health. In addition to this Executive Order, various federal laws relate specifically or are applicable to invasive species (see also ELI, 2002) (Table 1-1). The degree of focus on invasive species prevention, control, and management depends on the individual legislation.

Table 1-1. Summary of federal laws related to invasive species in aquatic ecosystems[a]

Legislation	Description of focus
Great Lakes Fish and Wildlife Restoration Act of 2006	Amends Great Lakes Fish and Wildlife Restoration Act of 1990 to further prevent intentional and unintentional introduction of sea lamprey.
Salt Cedar and Russian Olive Control Demonstration Act (2006)	Initiates assessment and demonstration program to control salt cedar and Russian olive.
Nutria Eradication and Control Act (2003)	Provides assistance to eradicate or control nutria and restore marshland.
Lacey Act of 1900; amended in 1998	Prohibits import of specific injurious species.
Federal Insecticide, Fungicide, and Rodenticide Act of 1996	Regulates importation and distribution of substances, including organisms, intended as pesticides, e.g., biocontrol agents.
National Invasive Species Act of 1996	Prevention focuses on ballast water introductions, especially to Great Lakes; authorizes research funding prevention and control; encourages coordination through regional panels; encourages state, interstate, and tribal invasive species management plans.
Nonindigenous Aquatic Nuisance Prevention and Control Act of 1990	Focuses on ballast water introductions to assess threats and control methods; establishes Aquatic Nuisance Species Task Force; establishes mechanism for approving state and regional aquatic nuisance species (ANS) management plans and funding ANS management activities.
Federal Noxious Weed Act of 1974	Bans importation and transportation of noxious weeds into or through U.S.

[a]Additional information available from http://www.invasivespeciesinfo.gov/laws/publiclaws.shtml.

Not all non-native species are harmful or will become invasive. For example, results from studies prepared for the former Congressional Office of Technology Assessment show that approximately 8% of terrestrial non-native species, 31% of non-native insects, and 28% of non-native fishes have had beneficial effects for humans and industry, such as species that provide biocontrol or serve as a foundation for agriculture (OTA, 1993). Further complicating the definitions of invasive species are those species that are considered damaging in one area but beneficial in another. The Asiatic clam (*Corbicula fluminea*) is one such example; it invaded a tidal marsh in the Potomac River in the late 1970s and increased water clarity to a level at which submerged aquatic vegetation reappeared and various aquatic bird populations returned to the area (Phelps, 1994). In other regions this AIS clogs intake pipes, damages industrial water systems, and alters habitat. Only a small percentage of non-native species become invasive and cause severe ecological and/or economic damage (OTA, 1993). For those species that do become invasive, their impacts can be devastating. Invasive species can threaten the very existence of native species in the invaded environments (Clavero and García-Berthou, 2005; Novacek and Cleland, 2001; Mack et al., 2000). Invasive species are a major cause of extinctions worldwide—48-62% of fish extinctions (68% of North American fish extinctions), 50% of bird extinctions, and 48% of mammal extinctions (Clavero and García-Berthou, 2005; Harrison and Stiassny, 1999; Miller et al., 1989). It should be noted that invasive species are generally not the only cause of these extinctions; invasive species are the sole cause for only 2 of the 40 extinct fish taxa in North America (Miller et al., 1989). In the U.S. alone, damage and losses from invasive species are estimated at approximately $120 billion annually (Pimentel et al., 2005). Also, despite advances in understanding what makes environments suitable for invasion and determining characteristics of species capable of invasion, it is still difficult to predict which species will become invasive (Richardson and Pyšek, 2006; Kolar and Lodge, 2001; Lonsdale 1999; Rejmánek and Richardson, 1996).

In this report we focus on AIS, including coastal, freshwater, wetland, and riparian species that are already problematic in one or more states and have the potential to expand into neighboring states as climatic conditions change. Species can also become invasive when introduced into areas with similar climates as their host climate, such as species from the Ponto-Caspian regions to the Great Lakes. Thus, we rely on the definition from Executive Order 13112 that the species has economic, ecological, or human health impacts to be considered invasive. However, because climate change has the potential to fundamentally change ecosystems, the way in which environmental managers differentiate and define native, non-native, and invasive species will also need to change in order to manage for changing threats.

AIS can cause a wide range of ecological impacts, including loss of native biodiversity, altered habitats, changes in water chemistry, altered biogeochemical processes, hydrological modifications, and altered food webs (Dukes and Mooney, 2004; Ehrenfeld, 2003; Findlay et al., 2003; Simon and Townsend, 2003; Eiswerth et al., 2000; Gordon, 1998). Wetlands, including estuaries, are some of the most invaded habitats in the world (Zedler and Kercher, 2004; Cohen and Carlton, 1998). Some of the most notorious U.S. invaders are aquatic species such as the zebra mussel, purple loosestrife, tamarisk, Asian carp, *Caulerpa* (marine green alga), and the green crab. Section 1.4 describes the ecological impacts of some of these invaders and the potential impacts of climate change on these species.

1.4. CLIMATE CHANGE AND ECOSYSTEM IMPACTS

The recently released Fourth Assessment Report Summary for Policymakers from the Intergovernmental Panel on Climate Change (IPCC) provides a comprehensive synthesis of the current state of climate change science and a discussion of the projected effects that climate change will have in the coming decades and centuries (IPCC, 2007). Atmospheric carbon has increased from 280 parts per million (ppm) in pre-industrial times to 379 ppm in 2005. Other greenhouse gases such as methane and nitrous oxide are also on the rise. Warming is occurring globally, as evidenced by increases in global mean air temperatures, global mean ocean temperatures, melting of snow and ice in polar regions and high altitudes, and sea level rise (IPCC, 2007). The projected effects of climate change include warmer and fewer cold days and nights over most land areas, warmer and more frequent hot days and nights over most land areas, increased frequency of warm spells and heat waves over most land areas, increased frequency of heavy precipitation events over most areas, increase in areas affected by drought, increase in intense tropical cyclone activity, and a rise in sea level (IPCC, 2007). Figure 1-1 presents the conceptual model of projected effects of climate change on aquatic ecosystems, interactions with AIS, and adaptive management responses. Some issues that are less well understood include how precipitation, groundwater recharge, and streamflow will change as a result of climate change (IPCC, 2001).

In addition to the physical changes, climate change is altering ecosystems and species life cycles. Changes include longer growing seasons in mid and high latitudes, shifts in species' ranges towards the poles and higher altitudes, decline of some species, and changes in the reproductive cycles of plants and animals that are cued by climate and seasons (Parmesan, 2006; Root et al., 2003; Walther et al., 2002; IPCC, 2001). In the U.S., species restricted to southern habitats may move north as milder winters allow overwintering. In other cases, less heat tolerant species may decline in their southern ranges, allowing for new species to fill the niches left behind (Aerts et al., 2006). Thermal lake stratification regimes may also be affected by warming water temperatures, resulting in earlier mixing and phytoplankton blooms that may alter zooplankton development. Changes to timing of zooplankton reproduction and/or abundance could favor certain species over others and have potential negative consequences for aquatic ecosystems (Winder and Schindler, 2004a, b). Altered hydrological regimes will also favor

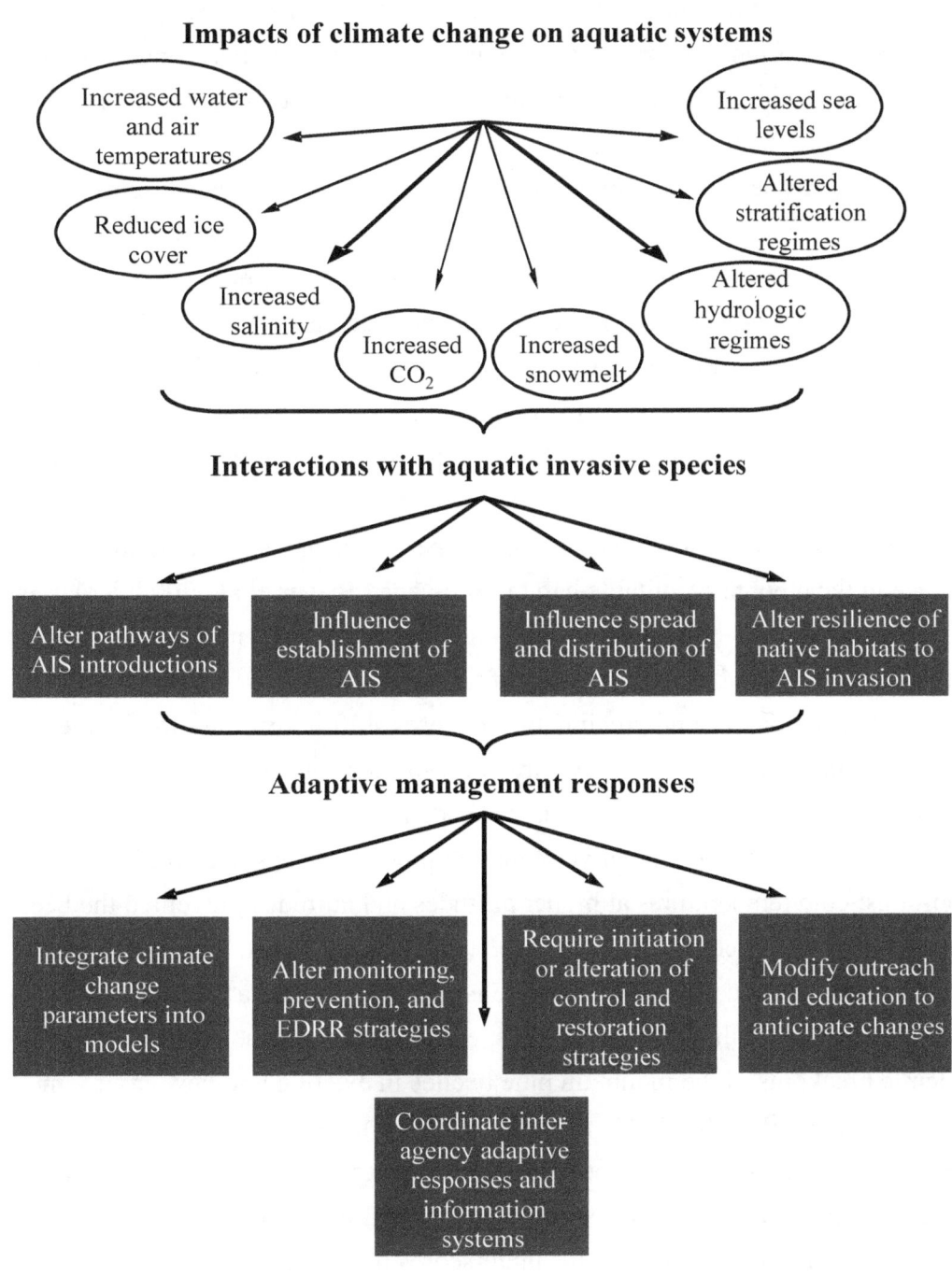

Figure 1-1. Climate change impacts on aquatic invasive species and possible adaptive management responses. Adapted from Rahel and Olden (2008).

some species over others. These changes may be particularly problematic for threatened and endangered species, the habitats of which are dwindling (McLaughlin et al., 2002), or those with limited dispersal capabilities if climate change makes their habitats unsuitable. Climate changes leading to increased rainfall or, conversely, drought may also shift invasive species ranges and present new opportunities for invasion. Climate change will also contribute to selective pressures on species, presumably leading to adaptive genetic changes that may influence species success (Barrett, 2000).

Some species do not require climate change to damage ecosystems, yet climate change may exacerbate the damage they do cause. Two examples of invasive species that alter the invaded ecosystem even without climate change are the common carp (*Cyprinus carpio*) and salt cedar (*Tamarix ramosissima*). The common carp decreases water quality and destroys viable habitat for other desirable species while the drought tolerant and deep-rooted salt cedar dominates riparian forests that were once dominated by cottonwoods and willows (Charles and Dukes, 2007; Kolar and Lodge, 2000; Lite and Stromberg, 2005). Climate change may have positive feedbacks for both of these invasive species if waters warm in the midwestern and northern U.S. and if the southwestern U.S. experiences more frequent droughts, leading to an increase in the amount of suitable habitat to invade (Seager et al., 2007; Kolar and Lodge, 2000). This interaction between climate change and invasive species may intensify ecosystem effects and possibly increase the spatial extent of these effects.

As temperatures and precipitation patterns shift in response to climate change, species ranges will also shift. A current example of a species shifting its range poleward and towards higher altitudes is that of the mountain pine beetle (*Dendroctonus ponderosae*). Historically, the range of the North American native mountain pine beetle has been limited due to climatic conditions; cold temperatures at higher altitudes and latitudes prevented the beetle from completing its life cycle in a single season (Logan and Powell, 2001). With increased warming at higher latitudes and altitudes the beetle is able to complete a life cycle in one season, allowing for range expansion, thus exposing new species of trees to pine beetle infestation and resulting in epidemic breakouts of the mountain pine beetles in existing and new environments (Carroll et al., 2003; Logan and Powell, 2001). Although this is a terrestrial example, it illustrates two important points: (1) invasive species are already responding to climate change and (2) native species can become invasive when they spread into new locations as a result climate change (Mueller and Hellmann, 2008). This underscores the importance of considering climate-change effects since species responses and overall impacts may not be limited to the current set of known AIS.

Currently, most examples of species' range expansions in response to climate change are terrestrial (see Parmesan, 2006; Root et al., 2003; Walther et al., 2002) although aquatic examples are increasing (Parmesan, 2006). An example of an aquatic plant expanding into higher altitudes is the threadleaf water-crowfoot (*Ranunculus trichophyllus*), which has invaded previously non-vegetated lakes in the Himalayas, an invasion attributed to climate change (Lacoul and Freedman, 2006). Tropical aquatic snails are another example of species whose ranges may expand under a changing climate. These snails are carriers of a specific genus of trematodes or blood flukes native to tropical and sub-tropical regions of the world that cause the disease schistosomiasis. These blood flukes could impact human health if these tropical aquatic snails, move northward as temperatures warm and conditions become more humid (Tol, 2002). While these aquatic species have not caused the types of damages attributed to mountain pine beetles, the potential exists for other AIS to cause further or unforeseen ecological or economic damages.

1.5. CLIMATE CHANGE IMPACTS ON INVASIVE SPECIES

Climate-change-induced alteration of ecosystem conditions can enable the spread of invasive species through both range expansion and creation of habitats and conditions suitable for newly introduced invasive species. Altered conditions such as increased atmospheric carbon dioxide, modified precipitation regimes, warming ocean and coastal currents, increased ambient temperature, and altered nitrogen distribution can increase invasive species success in some contexts (Ziska et al., 2007; Ziska, 2003a, b; McCarty, 2001; Dukes and Mooney, 1999). Research on climate change and invasive species is limited; however, many studies on potential climate-change impacts to aquatic systems and AIS exist, and several are highlighted here.

A number of scientific studies have examined whether increased atmospheric carbon dioxide may enable invasions; however, because most attempts to predict invasions have been on a small scale and knowledge of invasions is limited, predicting the effects of increased carbon dioxide is uncertain. The effects of carbon dioxide enrichment in aquatic ecosystems, especially with respect to AIS, are still much less well understood than in terrestrial ecosystems, with the exception of recent research on ocean acidification and consequences for coral reef ecosystems (Cao et al., 2007; Pelejero et al., 2005; Scavia et al., 2002). Research indicates that increased carbon dioxide in ocean and freshwater environments may alter macro- and micro- algae and plant dynamics (Feely et al., 2004). For example, Chen et al. (1994) found that increased carbon dioxide may cause the invasive aquatic plant dioecious hydrilla (*Hydrilla verticillata*) to increase its growth rate at elevated temperatures (e.g., maximum effects of temperature on growth were recorded at 25°C). Thus, as temperatures and carbon dioxide levels rise, hydrilla has the potential to spread more rapidly within and outside of its current range. Weltzin et al. (2003) examined how elevated carbon dioxide levels affect plant invasions in various ecosystems and concluded that increasing carbon dioxide levels will cause increases in resources, plant production, soil moisture, and nitrogen uptake, all of which create favorable invasion conditions.

Increased temperatures and altered precipitation regimes are likely to have larger effects on AIS than increasing levels of carbon dioxide. One study of emergent macrophytes in lakes showed that increased temperatures led to larger increases in biomass than increased carbon dioxide levels (Ojala et al., 2002). As in terrestrial environments, responses to carbon dioxide may be species specific, but other environmental variables like water temperature and hydrological regimes may be more important drivers in changing the establishment, spread, and impact of AIS.

Climate change is predicted to alter precipitation patterns, leading to droughts in some areas and flooding in others due to increased storm intensity. Knowledge of the effects of climate variability, which also causes droughts and floods, can offer some insights into how ecosystems respond to the stress of altered hydrology (Shafroth et al., 2002). There is much evidence in the invasive-species literature that ecosystem disturbances encourage pioneer species, and many invasive species are pioneers (Byers, 2002; Schnitzler and Muller, 1998). Thus, changes in precipitation due to climate change may affect AIS establishment and dispersal. Increased rainfall may allow for greater dispersal of upstream invasive species to downstream habitats. Zedler and Kercher (2004) hypothesize that wetlands are highly vulnerable to invasions because wetland invasive plant seeds are frequently dispersed by water. Lonsdale (1993) finds that flooding and rainfall are important factors affecting dispersal of the invasive weed *Mimosa pigra* in Australia. The size of the area colonized related to the amount of rainfall in the previous wet season, and the data suggest that seed dispersal by flotation is key to rapid wetlands expansion.

Increasing ocean temperatures also may enable new species invasions. Stachowicz et al. (2002) compare recorded sessile invertebrate species recruitment and establishment with temperature data. Their research shows that introduced ascidians (sea squirts) recruit earlier in years with warmer winter water temperatures while the recruitment of native ascidians did not significantly change with variation in winter water temperature. Because community composition is often determined by which species settles first, introduced ascidians may out-compete native ascidians as ocean temperatures warm. The authors also show that introduced ascidians have higher growth rates than native species at high temperatures. The authors conclude that rising mean winter water temperature is a stressor that may lead to increased invasions by non-native species in New England. In addition, as coastal currents warm, species may shift their ranges northward and become invasive in new areas. Barry et al. (1995) examined data on intertidal invertebrate assemblages in California using records that span over 60 years. These data show that near shore water temperatures increased by 0.75°C and summer temperatures increased by 2.2°C. Their study shows that southern invertebrate species increased in abundance and expanded their ranges while northern species that were not tolerant of warmer waters declined. Some of this response, however, may be due to cyclical tidal fluctuations that influence water temperatures (Denny and Paine, 1998). Overall, the change in species composition may be more species- and location-specific than a simple latitudinal response to temperature (Helmuth et al., 2006).

An example of a species that could move farther north as a result of warming waters is the invasive mussel, *Mytilus galloprovincialis*. This blue mussel species has a higher tolerance for warm water temperatures and increased salinity levels than the native blue mussel, *Mytilus trossulus*, in California. Thus, *M. galloprovincialis* has replaced the native mussel along much of the southern and central California coastline (Braby and Somero, 2006). The native mussel still dominates the northern coast of California as the invasive mussel is less tolerant of cold water temperatures (Braby and Somero, 2006). However, as temperatures warm in the North Pacific, *M. galloprovincialis* could have the potential to expand its range northward.

Dukes and Mooney (1999) also discuss increasing temperature in the context of climate change and find that it enables species invasions under certain circumstances. Mandrak (1989) examined 58 species to determine their potential to establish in the Great Lakes by comparing characteristics of these potential invaders to 11 recent invaders. Of the 58 species studied, 27 could potentially establish if temperatures warm from climate change, while the others could not. McFarland and Barko (1999) examined the effects of increased water temperature on a monoecious hydrilla, finding that the species is better adapted to higher temperatures than previously shown in the scientific literature. Populations of the common reed, *Phragmites australis*, also increase with higher-than-average ambient air temperatures (Wilcox et al., 2003). Another effect of warming temperatures may be an increase in the number of sexual versus asexual reproductive periods for plant species, resulting in increased rates of spread. Diaz-Amela et al. (2007) linked the flowering cycles of a Mediterranean seagrass (*Posidonia oceanica*) to warming water temperatures. If these types of changes occur in AIS, they may lead to further expansion and impacts.

1.6. INTERACTING GLOBAL CHANGE STRESSORS

Invasive species can be major ecosystem stressors, and their interaction with other global change stressors is not fully understood. Kolar and Lodge (2000) identify global change and other anthropogenic stressors that increase the number or the impact of freshwater invasive species: globalization of commerce (including shipping; bait, aquarium, and pond trade; and aquaculture); waterway engineering (including canals and dams); land use changes (including siltation, eutrophication, and water withdrawal); climate and atmospheric changes; and intentional stocking. Carlton (2000) identifies a slightly different set of global change and anthropogenic stressors affecting invasions in the oceans: overfishing, chemical pollution and eutrophication, habitat destruction and fragmentation, biological invasions (facilitating other invasions), and climate change. In the Great Lakes, human activities linked to aquatic invasions include clear-cutting and farming practices that increase sedimentation and water turbidity, industrial pollution, urbanization, and overfishing (Glassner-Shwayder, 2000). These examples show that there are many stressors interacting to facilitate the establishment and spread of invasive species and to influence the magnitude of their impact. Climate change will interact with existing stressors and may ameliorate or exacerbate their effects; however, little is known about the change in magnitude of effects due to climate change.

Although the above examples illustrate that there are many stressors interacting with invasive species and climate change, land-use and land-cover changes remain the major global stressors that affect these other stressors (Vitousek, 1994). Land-use change and the ecosystem disturbances it causes can lead to more invasions (Hansen and Clevenger, 2005; Mack et al., 2000). Nutrient loading due to increased agriculture, intensification of agriculture, or urban runoff can facilitate invasions of aquatic invasive plants (Lake and Leishman, 2004). Increased development can lead to degradation of habitats, and some studies demonstrate that degraded habitats are more prone to invasion than healthy environments (Mack et al., 2000). Hobbs (2000) discusses the complex nature of land-use changes and their effects on invasive species and habitat invasibility. Land-use changes include increased urbanization, deforestation, ecosystem fragmentation, and altered agricultural practices (intensification and abandonment).

Two additional major changes in recent history that can alter ecosystem dynamics are increasing levels of human transformation and domination of ecosystems (Vitousek et al., 1997b) and increasing transport of species leading to a breakdown of biogeographical barriers (Cohen and Carlton, 1998). Hobbs (2000) describes the complex interrelationship between land-use disturbances and invasions further. For example, land transformation (e.g., increased nutrient or pollution runoff from conversion to agriculture or urban development) can enhance invasion by providing opportunities for establishment. Invasion, in turn, can drive land transformation (e.g., an invasive tree species can convert grassland into forest). These processes may feed back upon each other to facilitate further alteration, possibly causing an "invasional meltdown," which leads to an acceleration in the number of invasive species and impacts (Ricciardi, 2001; Simberloff and Von Holle, 1999).

Climate change will present a major stressor with which managers and decision-makers will need to be concerned, particularly in the context of interacting with other contributors to species invasions. However, scientific understanding of the complexity of invasions resulting from climate change, and making predictions that incorporate this understanding, is not yet well developed. Indeed, one of the major challenges in investigating the interactive nature of climate change stressors is the incredible complexity of biological systems. Often, invasive species models account for climate change factors in isolation because of the challenges of complexity. For example, the most widely used models to predict invasibility may use temperature as the primary component, but in future modification and development of models, scientists should consider additional climate change parameters.

1.7. CLIMATE CHANGE AND INVASIVE SPECIES DISTRIBUTION MODELS

Models of invasive species introductions, distributions, spread, and establishment are key tools for both understanding the invasive-species problem and designing effective prevention and control techniques. Numerous types of models have been developed. In many cases, authors recommend that invasive-species managers be cognizant of specific factors (e.g., species interactions, climatic factors, spread vectors) in ecosystem management. Some offer clear, ready-to-use models and strategies for conservation managers. However, most models of species invasion currently do not explicitly account for climate change; this represents a research need that is discussed in sections 3.2.1 and 3.2.2, primarily in that scientists need to begin to build climate change variables and scenarios into these species invasion models. Initial steps include integrating climate change-related parameters such as salinity variations, temperature changes, and soil chemistry into these models and then examining future scenarios or projections. The outputs from models should help managers to better target and prioritize their prevention, monitoring, early detection, and rapid response programs under changing conditions. This section discusses some of the existing invasive-species distribution models and how climate-change information may be incorporated into them. Appendix E, Models for Invasive Species Introduction, Establishment, Spread, and Invasion provides additional examples.

1.7.1. Models to Assess Climate-Change Impacts on Species Distributions

Numerous ecological models have been developed to specifically address climate-change impacts on species distributions, but these models generally are not applied to invasive species. One of these types of models, the bioclimatic envelope model, is used to identify correlations between species' distributions and climate-change factors to determine a species' climatic boundaries. Based on this information, models predict how species' distributions may change under predicted climate changes (Pearson and Dawson, 2003). Discriminant analysis is one method that has been used to explicitly evaluate climate-change impacts on invasive species. Mandrak (1989) uses discriminant function and principal component analyses to compare ecological characteristics of possible invading species to recently invading species to determine potential invaders' response to climate change and found that nearly half of the potential species studied might invade in response to warmer temperatures. Curnutt (2000) used multiple discriminant analyses to identify connections between climate variables and plant distributions to predict plant invasions.

Ecological niche models also are used to predict potential species invasions. Several assumptions are fundamental to these models: (1) a species' distribution is limited by its ecological niche, and (2) a species can only disperse to an area with similar ecological characteristics (Peterson, 2003). One example of an ecological niche model is GARP (Genetic Algorithm for Rule-set Production), which can incorporate temperature as one of its environmental variables and has been used to predict invasive-species distributions (Kluza and McNyset, 2005; Peterson and Vieglais, 2001; Stockwell and Peters, 1999; Stockwell and Noble, 1992). Since temperature can be included as a predictor of species distributions, GARP can be modified to reveal the influences of changing temperature over time. Authors of several studies have used GARP to examine the potential effects of climate change on the distribution of species, including those of the invasive Argentine ant and *Limnopurna fortunei*, a freshwater mussel native to southeast Asia (Kluza and McNyset, 2005; Roura-Pascual et al., 2004; Peterson et al., 2002; Peterson and Vieglais, 2001). Underwood et al. (2004) developed a model using GARP to predict the environmental niches of non-native species in Yosemite Valley, California, using parameters of elevation, slope, and vegetation structure. Results demonstrate the predictive potential of GARP for identifying potential invasion sites. The study concludes that similar models can be developed for other national parks and that such models may increase efficiency of field work and monitoring and decrease cost to managers (Underwood et al., 2004).

Mechanistic approaches to modeling fundamental niches will provide additional predictive power to current models because these approaches identify how the characteristics that allow a species to survive (e.g., reproductive success, thermal tolerances, fitness requirements, and mechanisms for acquiring energy) interact with a species' biophysical environment. By also integrating spatial and temporal climate information, these models can provide a better landscape view of the elements of a species' fundamental niche as well as how species distribution and niche may change as climate changes. Climatic variables of specific niches may also be integrated into GIS maps to allow ecological managers to better visualize issues (Kearney and Porter, 2004; Porter et al., 2002).

The studies discussed above illustrate the potential usefulness of a variety of modeling approaches in projecting potential invasive species distributions under climate change. Challenges remain in integrating climate-change projections and mechanisms of invasion, particularly in aquatic ecosystems and in translating model results into information useful to managers and decision-makers.

1.7.2. Models to Assess Invasive Species Distributions

Most models that specifically address invasive species spread, distribution, and establishment do not incorporate climate change variables; however, many of these models could be modified to account for these changes. For example, diffusion models can be used to predict species dispersal patterns over a range of habitats (Buchan and Padilla, 1999; Grosholz, 1996). Factors that affect dispersion are important to the accuracy of these models; thus, they should incorporate climate-change factors, such as increased water temperatures and carbon dioxide and salinity levels, to determine how climate change may impact dispersal abilities and patterns. For example, boater movements facilitate zebra mussel dispersal (Buchan and Padilla, 1999). As temperatures stay warmer for longer periods of time and waterbodies remain ice-free for longer, boat traffic may increase and move into new areas. Diffusion models will need to account for these types of climate-induced changes in dispersal to ensure their accuracy.

2. MANAGEMENT OF AQUATIC INVASIVE SPECIES IN A CHANGING CLIMATE

2.1. STATE MANAGERS' REPORTED CLIMATE CHANGE CONCERNS

Each of the 50 states conducts management actions that address AIS problems. Programs and activities vary widely and may include research to assess current and future invasive threats or identify pathways; detection of newly established species (e.g., monitoring, surveys, inspection); import, introduction, or release requirements for species (e.g., permits and licenses); transport and shipping requirements; quarantine; education and public awareness efforts; control (e.g., biological, chemical, and manual); emergency response efforts; and restoration of degraded areas to increase resilience against re-invasion.

Many states have formed councils and developed management plans to organize and guide priorities for action and/or have dedicated funding for formal programs to address AIS problems. Other states conduct AIS management on a more ad hoc basis, under the purview of broader agency authorities. For example, a state parks agency might work to eradicate invasive species as part of the maintenance of a state-owned recreational area; a state wildlife agency might seek to protect regulated fish and game species by preventing or controlling invasive threats. In any case, each of the 50 states, albeit to varying degrees, performs some form of AIS management.

In order to determine the information needed to allow state AIS managers to consider and incorporate projected climate change effects into their programs, we inventoried AIS-related management actions in all 50 states (see Appendix A: Aquatic Invasive Species Programs and Activities). Research entailed the review of publicly available documents, publications, and online materials. For further clarification when appropriate, we discussed AIS programs, research needs, and management strategies with AIS managers, scientists, and decision-makers. Discussions during two workshops organized as a part of this effort also contributed to the information on climate-change concerns.

Results suggest that many managers and decision-makers are cognizant of the potential impacts of climate change on invasive species and the effect this driver may have on the goals and objectives associated with existing activities and decisions. Reported concerns emphasize how climate change will exacerbate existing problems, and how it may enhance conditions suitable for invasive species not previously established. Following is a list of concerns reported by states:

- AIS range expansions

- Identification of species that are more likely to establish under changing conditions and modification of management priorities, accordingly

- Prediction and assessment of conditions that may lead to invasion (e.g., warmer temperatures, disturbed ecosystems and native species, increased nutrient availability, modified precipitation regimes, and erratic weather patterns)

- Overwintering capabilities for invasive species

- Increased propagule pressure and vectors

- Increased growth rates

- Unanticipated interactions between climate change and invasive species

- Effects of climate change on the success of control efforts

- Effects on ecosystem services from increased invasions (e.g., water supply, recreation, etc.)

While state management staff generally recognizes that climate change is an important issue, most states have not begun to incorporate climate-change information into their ongoing AIS programs, activities, or plans; few programs make concrete decisions based upon projected climate-change impacts. However, there is no specific guidance directing the inclusion of climate-change considerations in state AIS programs or plans. Additional challenges not reported by states—which may also highlight the nascence of the issue for many state managers—include, among others, the potential effects of changes in climate on control methodologies and costs, management and authority, and communication of the problem to the public.

Although not every state operates a comprehensive AIS program, consideration of the effects of climate change is still essential to the success of those management efforts that are undertaken. Because states' resources for invasive-species management are often scarce, they should be used to support management activities that will prevent, control, and eradicate species in as efficient a manner as possible. Incorporating climate-change information when planning and implementing prevention, control, and eradication activities will help maintain the manager's ability to successfully carry out these activities. Adopting an adaptive management framework for AIS management practices will allow states to be better equipped to prevent and control AIS invasions under changing conditions and will also maximize the effectiveness and efficiency of each state dollar spent on such activities.

2.2. MANAGEMENT PLANS AS BLUEPRINTS FOR ACTION

Congress passed the Nonindigenous Aquatic Nuisance Prevention and Control Act of 1990 (NANPCA) to address the national problem of AIS. Section 1204 of NANPCA allows governors to submit management plans that identify areas and activities that would benefit from technical, enforcement, or financial assistance in order to eliminate or reduce the environmental, public health, and safety risks associated with AIS. Once these management plans are approved by the Aquatic Nuisance Species Task Force, states or regions are eligible to receive federal funding to assist with prevention and control activities. To date, 23 state AIS plans have been approved by the Task Force and several states have plans in various stages of development (MacLean, 2007). In FY 2006, Congress appropriated more than $1,075,000 of cost-share funding for states to implement their plans.

Management plans are often organized into the following categories of action:

- Leadership and coordination
- Prevention
- Early detection and rapid response (EDRR)
- Control and management
- Restoration
- Research
- Information management
- Education and public awareness

In addition to inventorying AIS-related management actions for all 50 states (see Appendix A, Aquatic Invasive Species Programs and Activities), we also reviewed completed state and regional AIS management plans available at the time of the study and assessed how they incorporate climate-change considerations specifically, as well as how they provide for adaptation of strategies and actions under changing conditions more generally (see Appendix B, State Aquatic Invasive Species Management Plan Summaries and Appendix C, Regional Aquatic Invasive Species Management Plan Summaries). Regional plans generally serve to coordinate activities among states and their AIS management plans, while state plans outline more specific activities. Existing plans are in various stages of both development and implementation, and some states operate a multitude of AIS management activities and programs in the absence of a plan. However, an assessment of state plans provides a logical starting point for understanding how states are anticipating and responding to predicted effects of changes in climate.

2.3. RESEARCH RESULTS: STATE PLANNING FOR CLIMATE CHANGE AND AQUATIC INVASIVE SPECIES

In total, we reviewed 25 state plans (approved plans, plans pending approval, and draft plans), including 23 AIS-specific plans and two general invasive species management plans with a significant AIS focus. Several other states are currently developing AIS management plans, which we did not include in our review. We were also unable to include several recently approved state plans as they were not available at the time of analysis, including California's, Idaho's, and Rhode Island's plans. In addition to the 25 state plans, we reviewed seven regional AIS plans.

Table 2-1 summarizes how each state's plan (1) addresses potential impacts resulting from climate change, (2) demonstrates capacity to adapt goals and activities to changing conditions, (3) provides monitoring strategies, (4) includes plans for periodic revision and update of the plan, and (5) describes funding sources/strategies for plan implementation. Within each category, several more specific questions were examined (see Appendix D for the list of questions under each category). For each question we assigned scores from 0 to 3. A score of 0 meant that a plan had no evidence of capacity to address a particular question or set of activities. A score between 1 and 3 meant there was some level of capacity or potential for that state to incorporate and address information on and impacts from changing conditions, including climate change.

Our assessment revealed that few plans incorporate climate change or the resultant change of conditions (see Appendix D, Complete Criteria and Scoring for State Plan Consideration of Climate Change and/or Changing Conditions for the full criteria and scoring). The majority of state plans have management actions that, if conducted under different environmental conditions, may prove less relevant, less efficient, or less successful than they are under current conditions. However, some states, such as Alaska, Hawaii, and Washington, recognize that conditions may change over time and have built considerations of changing conditions into their management actions. In addition, many state plans contain measures to periodically review and update management strategies and tasks, providing the opportunity to review the robustness of management plans in light of climate change and to amend plans where feasible.

While most state plans do not mention climate change or changing conditions, our assessment of these plans does reveal that states have some capacity to adapt their program or activities (Table 2-1). The assessment results represent a potential adaptive capacity across different parts of each of the state's program, which should make it easier for managers and decision-makers to address potential program vulnerabilities to climate change. The scores from answering each of the questions within each of the five categories assessed (impacts from

Table 2-1. Consideration of climate change and/or provision for adaptation of strategies and actions under changing conditions in 25 AIS management plans*. Possible total score is 54.

	Understanding and incorporating potential impacts resulting from climate change (out of 15 total points)	Capacity to adapt goals and activities (out of 24 total points)	Monitoring strategies (out of 9 total points)	Plan includes strategy for updating and incorporating new information (out of 3 total points)	Plan identifies dedicated funding source for implementation (out of 3 total points)	Score
Washington	3	3	6	3	2	17
Alaska	4	4	5	2	1	16
Hawaii	4	3	6	1	0	14
Kansas	0	3	6	3	2	14
Connecticut	3	4	2	1	2	12
Indiana	3	2	3	3	1	12
Louisiana	6	1	3	0	2	12
Missouri	3	0	6	0	3	12
Massachusetts	5	0	3	0	2	10
Maine	5	0	1	3	0	9
Montana	1	3	0	3	2	9
North Dakota	3	1	2	1	2	9
Oregon	3	0	0	3	3	9
Iowa	1	0	3	2	2	8
Wisconsin	1	0	3	1	3	8
Virginia	4	0	0	3	0	7
Arizona	1	2	3	0	0	6
Illinois	2	2	2	0	0	6
South Carolina	0	0	1	2	3	6
Ohio	1	2	1	1	0	5
Texas	0	0	5	0	0	5
Michigan	1	0	3	0	0	4
New York	0	0	3	0	1	4
Pennsylvania	0	0	0	2	1	3
Idaho	1	0	0	0	0	1

*To view the complete set of criteria and scoring for each state, see Appendix D, Complete Criteria and Scoring for State Plan Consideration of Climate Change and/or Changing Conditions.

climate change, capacity to adapt goals and activities, monitoring strategies, plan updates, and plan funding) were summed to give the category scores listed in Table 2-1. Scores for each question could range from 0 to 3, such that any score above 0 meant that some capacity existed and the difference between a score of 1 and 3 was the degree to which this capacity was explicitly acknowledged. Most states (92%) scored at least 1 in one or more of the five categories assessed. These results also illustrate which aspects of state programs can be modified more readily. For example, when scores are summed across states for each category and normalized by the number of questions assessed in each category, most of the adaptive capacity is in two categories, plan updates and plan funding. The ability of plans to be revised to incorporate new information and the fact that states have sources of funding to accomplish goals and activities show that plans could incorporate climate-change information and take steps to modify goals and activities. Monitoring strategies is the next category where state plans exhibit substantial adaptive capacity. Relative to the other four categories, the category describing goals and activities currently shows the least amount of adaptive capacity.

The highest scoring state was Washington with 17 points, though that still is less than half of the possible 54 points. This plan has the highest score in our assessment in part because it includes statements on the effects of climate on species boundaries, has a specific plan for using, managing, and updating monitoring data, and includes a timeline or benchmarks for updating the plan with new information. For example, Washington's plan includes information on various invasive species such as giant salvinia (*Salvinia molesta*) and water hyacinth (*Eichornia crassipes*) that have temperature ranges that currently prevent them from invading the state—information important for modifying prevention and monitoring activities in response to climate change (Bierwagen et al., 2008; Meacham, 2001) (for detailed results on each state plan, see Appendix B, State Aquatic Invasive Species Management Plan Summaries).

2.3.1. Understanding and Incorporating Potential Impacts Resulting from Climate Change

Only the Virginia state AIS management plan includes a discussion of climate change. Overall, 84% of the plans assessed fail to even mention climate change. However, most state plans (76%) include information on temperature tolerances of species, and some (40%) include discussions about the sensitivity of ecosystems to changing conditions. These results indicate areas where capacity exists in most states to begin to identify how these species may respond as climate changes at their current boundaries. None of the reviewed plans identify climate change effects as potentially important research topics or mention the regional differences in projected climate changes.

2.3.2. Capacity to Adapt Goals and Activities to Changing Conditions

Table 2-2 provides an assessment of the capacity of each state's plan – as it is currently constituted – to adapt its goals and activities related to leadership and coordination, prevention, EDRR, restoration, research, information management, and education and public awareness to changing conditions. Fewer than half of state plans (48%) mention changing conditions, and this is generally implicit in the types of goals and strategies described that could be used to respond to any changes in the environment, including climate change. No state plan that was examined accounts for changing conditions in its restoration or information management goals and strategies—two critical aspects of a comprehensive AIS management plan—though many plans do express the need for research and data to inform management decisions under changing conditions in the states' research goals and strategies. Of the plans that mention changing conditions under 'Research,' 20% explicitly mention research into changing conditions (scores of 2 or 3). Counting both implicit and explicit mention of changing conditions in these categories shows slightly higher capacities across states although the research category still dominates (Table 2-3).

The goals and activities described by state plans in each of these sub-categories are likely to be affected by climate change. For example, prevention activities will be challenged as species move outside of known ranges. Modifications to how vectors and pathways are monitored may be necessary to capture these effects. One approach may be integrated vector management (Carlton and Ruiz, 2005). The integrated vector management framework distinguishes cause, route, and vector for an invasion, including the biological and anthropogenic dimensions. This breakdown into the components is useful for analyzing where climate change may interact with vectors in order to formulate appropriate management responses.

2.3.3. Monitoring Strategies

Although no plan includes a specific strategy for monitoring changing environmental conditions, Maine's plan does note that climate change could be a cause for potential spread of AIS and that the state will monitor climate conditions to provide early warning of new populations (Dominie Consulting and Maine Interagency Task Force on Invasive Aquatic Plants and Nuisance Species Technical Subcommittee, 2002). Most plans, however, do have clear strategies for using, managing, and updating monitoring data (80%). These results show a high capacity to modify activities associated with monitoring to include information on climate change effects.

While new information on more effective management methods can be incorporated into many of the plans, climate change may pose additional challenges with respect to the spatial and temporal scales of monitoring (Hellmann et al., 2008). Feedback from researchers to managers

Table 2-2. Breakdown of which sections in AIS management plans account for changing conditions in goals and activities.[a] Possible total score is 24.

| | Sections in plans that account for changing conditions in goals and activities related to...[b] | | | | | | | |
	...leadership and coordination	...prevention	...EDRR	...control and management	...restoration	...research	...information management	...education and public awareness
Alaska	1	1	0	0	0	1	0	1
Arizona	0	1	0	0	0	1	0	0
Connecticut	1	1	0	1	0	1	0	0
Hawaii	0	0	1	0	0	1	0	1
Idaho	0	0	0	0	0	0	0	0
Illinois	0	0	0	0	0	2	0	0
Indiana	0	1	1	0	0	0	0	0
Iowa	0	0	0	0	0	0	0	0
Kansas	0	0	0	0	0	3	0	0
Louisiana	0	0	1	0	0	0	0	0
Maine	0	0	0	0	0	0	0	0
Massachusetts	0	0	0	0	0	0	0	0
Michigan	0	0	0	0	0	0	0	0
Missouri	0	0	0	0	0	0	0	0
Montana	0	0	0	0	0	3	0	0
New York	0	0	0	0	0	0	0	0
North Dakota	0	0	0	0	0	1	0	0
Ohio	0	0	0	0	0	2	0	0
Oregon	0	0	0	0	0	0	0	0
Pennsylvania	0	0	0	0	0	0	0	0
South Carolina	0	0	0	0	0	0	0	0
Texas	0	0	0	0	0	0	0	0
Virginia	0	0	0	0	0	0	0	0
Washington	0	0	0	0	0	2	0	1
Wisconsin	0	0	0	0	0	0	0	0

[a]To view the complete set of criteria and scoring for each state, see Appendix D.

[b]Scoring: 0 = none; 1 = implicitly (i.e., includes goals and strategies that can be used to account for changing conditions, but does not specify changing conditions as part of their purpose); 2 = yes, explicitly, in passing; 3 = yes, explicitly, and specifies associated goals and/or action items.

Table 2-3. Percent of plans implicitly or explicitly accounting for changing conditions

Plan chapter	Percent of plans
Leadership and coordination	8%
Prevention	16%
EDRR	12%
Control and management	4%
Restoration	0%
Research	40%
Information management	0%
Education and public awareness	12%

about changing conditions would be valuable in order to adapt management activities. Thus, regional coordination, links between research and implementation, and decisions about the scale of monitoring could be included in invasive species management plans to build on their existing capacity.

2.3.4. Plan Revisions and Funding

Most of the state plans include language about periodic revisions (64%), which indicates a high capacity to include new information and update goals and activities. Hawaii's plan even recommends a regular update specifically to address and adapt to changing conditions (Shluker, 2003). Thus, state plan revisions may include information about climate change effects in the future. Although only 16% of states reviewed specify a source for 100% of the required funding for their actions (i.e., Missouri, Oregon, South Carolina, and Wisconsin), most state plans do identify some funding associated with their goals and activities (64%). This identification of funding indicates an overall high capacity for states to accomplish tasks in management plans. Combined with periodic revisions, the allotment of funding demonstrates that many of these states could accomplish activities that may ameliorate climate-change effects on their invasive species programs.

2.3.5. Adaptive Capacity in Regional Plans

Table 2-4 gives an overall score for how each regional plan addresses climate change and demonstrates a capacity to adapt to changing condition. The plans were scored using the same scoring system set up for state AIS management plans and shown Table 2-1. One of the seven regional AIS management plans, the Lake Champlain Aquatic Nuisance Species Management

Plan, shows significant capacity to adapt to changing conditions. The plan scored significantly higher than the other six regional plans, a score that reflects the inclusion of a strategy for updating and managing monitoring data as well as regularly updating the plan. In addition, action items for EDRR, control, and management implicitly include climate-change considerations. Other regions do not exhibit the same capacity; thus, incorporating information on changing conditions, climate-change effects, and possible adaptive responses could greatly increase the adaptive capacity of regional management plans.

Table 2-4. How seven regional plans account for changing conditions. Possible total score is 54.

Regional Plan	Score
Lake Champlain Basin Aquatic Nuisance Species Management Plan	12
Midwest Region Aquatic Nuisance Species Action Plan	3
Gulf of Mexico Aquatic Nuisance Species in the Gulf of Mexico: A Guide for Future Action by the Gulf of Mexico Regional Panel and the Gulf States	1
Southeast Region Aquatic Nuisance Species Action Plan	1
Western Region Aquatic Nuisance Species Action Plan	1
Northeast Region Aquatic Nuisance Species Action Plan	0
Great Lakes Action Plan for the Prevention and Control of Nonindigenous Aquatic Nuisance Species	0

2.3.6. Conclusions about Adaptive Capacity as Illustrated in State and Regional Plans

Our examination of 25 state plans' and seven regional plans' capacities to adapt to changing conditions shows that few states and regions have developed strategies and associated tasks that specifically address climate change or consider potential changes in environmental conditions in general. While this is not a surprising finding, since states and regions currently are not mandated to consider climate-change effects and have limited resources for AIS management activities, management plans could incorporate more strategies to increase a state's or region's capacity to adapt to changing conditions. This analysis highlights that some capacity exists to deal with the additional stressor of climate change, particularly through revisions of management plans, the ability to fund specific activities, and the existing monitoring strategies. These results provide managers and decision-makers with information on what aspects of management plans can be readily revised to incorporate climate-change information and where adaptive management approaches may be most beneficial.

The following sections summarize how state AIS management activities, including leadership and coordination, prevention, control, restoration, and information management, may be adapted to address the predicted effects of climate change. The options presented are intended as examples that managers and decision-makers can consider when modifying AIS management plans to incorporate effects due to climate change. To learn more about specific, individual state and regional AIS management plans and how they can be revised to incorporate climate considerations and adaptive management procedures, see Appendix B, State Aquatic Invasive Species Management Plan Summaries, and Appendix C, Regional Aquatic Invasive Species Management Plan Summaries.

2.4. ADAPTING STATE PROGRAMS, ACTIVITIES, AND PLANS TO INCORPORATE CLIMATE CHANGE CONSIDERATIONS

In the sections below, we discuss how state programs, activities, and planned action items related to each category of activity may be susceptible to the projected effects of climate change, and we make recommendations for how management plans and strategies could be adapted to account for and remain robust under changing conditions.

2.4.1. Adapting Leadership and Coordination Activities

Coordination among federal, state, and local agencies, conservation organizations, and key members of the private sector allows for comprehensive and complementary coverage and implementation of state AIS plans and programs, as well as more efficient identification of priority issues and concerns (ELI, 2002). To facilitate coordination and provide leadership on AIS issues, many states have established invasive species councils, working groups, or task forces, and other states have invasive plant-, pest-, or AIS-specific councils (NISIC, 2006). Many states have also hired state agency staff to coordinate state (or agency) management tasks among agencies, conservation organizations, landowners, and other stakeholders. Finally, state AIS plans, often created under the leadership of a state council, play a fundamental role in guiding state AIS management strategies and management actions.

As leaders in AIS management, invasive species councils at state and regional levels are in an excellent position to begin to address climate change. Councils may consider holding meetings or workshops to (1) understand the scope of the climate-change problem and its potential effects on AIS; (2) modify the design, if necessary, of current management actions and plans to incorporate existing climate-change information; and (3) identify further informational and leadership needs. For example, states and regions will need to know

- how environmental conditions may change;

- which species may become threats under projected future conditions, including temperature tolerances of species;

- which systems may become vulnerable to invasion due to changes in temperature, nutrient availability, water quality or quantity, and/or changes in ecological community composition;

- how vectors will be influenced by changes in climate;

- how management actions, such as control methods, may be affected by changes in the environment; and

- what research is needed to better inform management strategies.

State councils (or agency staff in the absence of a council) would benefit from sharing climate-related concerns and data with other states to address regional species of concern due to shifts in climate. For example, neighboring states could be alerted to encroaching species, changing vectors, and modified control strategies when possible. Lists of potential invaders could be created and distributed among neighboring states when possible. Regional councils may also be useful in coordinating these activities.

State councils may also play a role in coordinating cross-program integration for strategies and tasks that involve more than one state agency or more than one division within agencies, particularly those aspects that may involve multiple media.

Not every state will have the resources to develop an organized, systematic approach to address climate change. In these states, natural resource management and environmental protection agency staff and coordinators may begin incorporating climate-change information by reviewing current prevention, control, and eradication activities, as well as planned action items, for their potential vulnerability to climate change; identifying information needs; and modifying strategies where feasible and when climate information is available from the growing body of related literature or from knowledgeable practitioners and researchers.

2.4.2. Adapting Prevention Activities

Prevention measures are implemented to avoid the introduction and establishment of invasive species and are widely recognized as the most effective and cost-efficient tools for combating invasive species (Keller et al., 2007; Leung et al., 2002; NISC, 2001; Wittenberg and Cock, 2001). Addressing invasive species through prevention mechanisms such as early detection and eradication will be less costly over the long-term than post-entry maintenance and control activities that depend on continued commitment and resources as well as on development of successful, targeted control mechanisms (Simberloff, 2003; Mack et al., 2000). Focusing on post-entry activities also can result in a significant time lag between identification of the invasion and implementation of control mechanisms—a delay that could result in extensive spread and establishment of invasive species (Ruiz and Carlton, 2003). In addition, increased invasions by numerous different species mean that various invasive-species management techniques are needed, and with limited resources, only a fraction of these management actions can be implemented (Ruiz and Carlton, 2003). Thus, prevention activities are key tools for successfully addressing invasive species, and states with limited resources may maximize the use of scarce invasive-species dollars by investing in prevention efforts.

Numerous strategies and measures may be used to prevent the establishment of potentially harmful AIS, including mapping and/or surveys to identify and mitigate invasive species threats, regulation of certain species (e.g., introduction, import, or release requirements), quarantines, EDRR protocols and emergency powers to quickly identify and address new infestations, and education to increase public awareness regarding particular species and/or pathways. Another important prevention tool for invasive species managers is the Hazard Analysis and Critical Control Points (HACCP) planning framework. As a part of the HACCP planning process, natural resource managers identify potential invasive species and possible points of entry that could result from management activities. Managers also focus on specific pathways and develop best management practices to prevent these species from being introduced. This planning framework helps managers assess risk and make more strategic decisions (USFWS, 2005).

Many state AIS prevention efforts are specific to species that have been identified as imminent threats` while other activities are focused on managing and responding to common AIS pathways such as ballast water, recreational boating, water gardening, or aquaculture. For example, the New Hampshire Department of Environmental Service's Weed Watcher Program trains volunteers to inspect recreational boats and other recreation-related gear to prevent introduction of aquatic invasive plants. Often, states will conduct a combination of prevention measures to address species or pathways. The Maryland Department of Natural Resources - Fisheries Service, for example, seeks to prevent the spread of snakeheads by circulating posters that ask anglers to kill and report all snakeheads, compiling regional data for captures in the Potomac River, and annual monitoring that includes seine, electrofishing, and gillnet surveys. Maine's Department of Environmental Protection and Department of Inland Fisheries and Wildlife conduct aquatic invasive plant prevention along common pathways. The agencies jointly inspect watercraft, trailers, and outboard motors at or near the state borders and at boat launching sites, regularly patrol waters and roads, and enforce violations such as launching a boat or transporting a vehicle on public roads with plants attached.

Prevention activities typically focus on species that are already known to cause impacts. Climate change, however, may enhance environmental conditions for some species with the following consequences: (1) new species are now able to survive in these locations, (2) known invasive species expand their range into new territories, and (3) species that are currently not considered invasive may become invasive and cause significant impacts. Monitoring and survey efforts may be used to identify species that are encroaching as a result of expanding ranges. Monitoring efforts, as given in Figure 1-1, may need to be modified to focus on weakened or changing ecosystems that are more vulnerable to invasion (Hellmann et al., 2008). As temperatures warm, precipitation regimes fluctuate, and nutrient flows change, ecosystems may lose their ability to support a diverse set of native species, becoming more vulnerable to invasion as new resources become available; however, managers should not assume that pristine, species-rich environments are immune to invasion (Melbourne et al., 2007; Byers and Noonburg, 2003; Davis et al., 2000).

Vectors also may be influenced by changes in climate and should be evaluated for their ability to transmit species under changing conditions. For example, seaways may remain open for longer periods during the year due to warming temperatures; thus, shipping and boating traffic, a major vector for species such as the zebra mussel, also may increase. In addition, completely new shipping routes may also open in polar waters due to melting ice, which will further increase boating and shipping traffic (Hellmann et al., 2008; Pyke et al., 2008). To begin to address these concerns, pathway analysis and species prediction models should be modified to include climate change parameters. States may need to alert inspection and border control agencies to new invasive threats, and related inspection priorities may need to be re-assessed in light of these impending threats and pathways. Import/introduction/release requirements should be based on risk assessments that account for how changing conditions will affect the potential for an area to be invaded. Climate changes resulting in increased storm surge and flooding may increase the risk of species escape from aquaculture facilities. In light of these changes, aquaculture facilities may need to take additional precautionary measures against escapes or establishment (e.g., use only triploids, stock only one sex, or use sterile hybrids) or to use only native species. Finally, ongoing land and water management activities should be re-evaluated for their potential to provide new invasion pathways. For example, waterway engineers could examine passage between water bodies that were historically separated, create barriers to passages, and consider AIS spread before re-filling or reconnecting waterways (Rahel and Olden, 2008).

2.4.3. Adapting Early Detection and Rapid Response Activities

EDRR refers to efforts that identify and control or eradicate new infestations before they reach severe levels. Because even the most effective barriers to entry will at some point be breached, EDRR is an important element in preventing and controlling invasive species problems. In addition to surveying and/or mapping to detect infestations, EDRR efforts may include emergency powers for state agencies to implement control measures quickly and restoration to decrease vulnerability to re-establishment of the invading species. Comprehensive EDRR plans identify participating and lead agencies, potential regulatory requirements for control, and other EDRR protocols.

The effectiveness of EDRR efforts may be improved by monitoring both for the establishment of new infestations as well as for changing conditions in order to better predict which systems may become vulnerable to invasion. To address the potential effects of climate change, continued and new monitoring will be necessary to update information systems with data that allow evaluation of those effects (Lee et al., 2008). Adapting monitoring may mean sampling at different temporal or spatial frequencies, or using different sampling techniques (Hellmann et al., 2008). For example, monitoring to detect range changes may require sampling the distributional and altitudinal edges of species ranges.

2.4.4. Adapting Control and Management Activities

Control and management measures vary widely among states and depend on the species being targeted, the infested ecosystem, availability of resources, and severity of the infestation, among other factors. Control techniques may be biological, chemical, manual, or mechanical, or a combination thereof. EDRR is an important element of an overall invasive species control strategy (see Section 2.4.3. Adapting EDRR Activities).

Changing conditions, such as warmer waters, extreme weather events, salt water intrusion, and/or changes in water chemistry, may affect the success of "tried and true" biological, chemical, or mechanical control measures. To guard against ineffective control measures, managers must be aware of the conditions under which an introduced biocontrol species may fail—or conditions under which they may thrive beyond control—and cross-reference those parameters with projected changes in the ecosystem. Changes in temperature and precipitation may affect biocontrol and invasive species differently, either increasing or decreasing the effectiveness of the biocontrol agent (van Asch and Visser, 2007; Stireman et al., 2005; Bryant et al., 2002). For example, salt cedar leaf beetles (*Diorhabda elongate*) may be less effective at controlling tamarisk (*Tamarix ramosissima*) in warmer temperatures, while the alligatorweed flea beetle (*Agasicles hygrophila*) may become more effective in controlling alligatorweed (*Alternanthera philoxeroides*). Similarly, herbicides and other chemical control measures may also be affected by temperature, water chemistry, and other climate-related changes in the ecosystem (Ziska et al., 1999). Finally, mechanical control may no longer be feasible when warmer winter temperatures allow invasive species to spread that are currently limited by hard freezes or ice cover and occur in limited areas. A re-evaluation of appropriate control measures may be necessary in order to make efficient use of state investments in AIS management.

2.4.5. Adapting Restoration Activities

Restoration of natural systems is critical to preventing re-introduction of an invasive species once it has been eradicated or controlled. Because healthy ecosystems can be less vulnerable to invasion (Vitousek et al., 1996), restored ecosystems also may be less vulnerable to future invasions, thus providing some insurance to investments in invasive species prevention, EDRR, and other control measures. One example of this use of restoration is Massachusetts' Aquatic Invasive Species Management Plan that calls for reintroducing native species as part of a restoration program for lakes and ponds (Massachusetts Aquatic Invasive Species Working Group, 2002).

Given that climate change is expected to alter native species and habitats and other ecosystem attributes, restoration designs should emphasize restoration of ecosystem processes (e.g., sediment and nutrient transport, export of woody debris, river-floodplain connections) that were originally disrupted and may have facilitated the establishment of AIS. Restoration projects should include analyses of which native species may thrive in, or at least tolerate, future climate-change conditions and avoid those species that may not be as well suited to future conditions. Restoration plans that include the effects of sea level rise and the increased occurrence of extreme weather events are likely to produce projects that remain effective under future climates. For example, state coastal restorations are expected to be at risk from climate change because water levels are critical in marsh restorations, and sea level rise could render many current saltwater marsh restorations useless if this effect is not considered in plans. Based on these factors, states may modify long-term restoration strategies in order to make habitats more robust and less vulnerable to potential invasions as conditions change.

2.4.6. Adapting Information Management Activities

No state has adopted a formal information management system that documents, evaluates, and monitors impacts from invasive species (NISC, 2001). State agencies that are considering the development of an information management system will have to support rapid and accurate discovery of data, correlate and synthesize data from many sources, and present the results of data synthesis that meets the needs of users. In addition to data on species movement and establishment, information on ecosystem conditions—e.g., water temperatures, chemical composition, and salinity levels, where applicable—should also be monitored and evaluated to fully assess invasive-species threats in the context of a changing climate. Any existing or planned information systems for AIS should incorporate information on climate change and its effects on invasive species (Figure 1-1) and have the ability to be updated with monitoring information in order to assess the occurrence of effects (Lee et al., 2008). As more information on effects of climate change on AIS becomes available, information systems will require the capacity to be updated. Then more targeted research may be done that can provide more specific recommendations for AIS management in a changing climate (see also Section 3).

2.4.7. Adapting Public Education Activities

Many states conduct public awareness campaigns to inform the public, decision-makers, and other stakeholders about ways to prevent the introduction and spread of invasive species. For example, Nevada's Lake Tahoe Basin Weed Coordinating Group posts signs and distributes information to boaters on boat cleaning and disseminates flyers to alert them about potential AIS spread. Similarly, the Utah Department of Natural Resources' Division of Parks and Recreation and Division of Wildlife Resources educate boat drivers from areas of known zebra mussel infestations, encourage and fund boat washing, and inspect boats for infestations. The programs also post public alert signs at major recreational waters, include AIS-information inserts in boat re-licensing packets, and print and distribute AIS brochures.

Modifying outreach and education efforts to incorporate information about climate change effects on AIS and their management is another possible management response presented in Figure 1-1. State AIS outreach campaigns can use their existing efforts to educate the public about new invasive species threats due to climate change.

2.5. EXAMPLE MANAGEMENT RESPONSES TO CLIMATE CHANGE

States conduct management activities that target a wide variety of AIS. Based on a review of the state AIS management programs (see Appendix A, Aquatic Invasive Species Programs and Activities: 50-State Summary), the following species were commonly reported as problems (five or more states reported the species as a problem): Asian carp species such as grass carp (*Ctenopharyngodon idella*), zebra mussel (*Dreissena polymorpha*), water hyacinth (*Eichhornia crassipes*), hydrilla (*Hydrilla verticillata*), purple loosestrife (*Lythrum salicaria*), Eurasian water milfoil (*Myriophyllum spicatum*), common reed (*Phragmites australis*), curly leaf pondweed (*Potamogeton crispus*), giant salvinia (*Salvinia molesta*), salt cedar (*Tamarix ramosissima*), and water chestnut (*Trapa natans*).

This section discusses four common AIS that are current priorities for many states, including one marine species, and examines how climate change may affect these species. Although many other species are also high priorities, many of the management activities and potential responses to climate change may be transferable from these examples. Each example illustrates how climate change can both positively and negatively affect current management and control activities. Where the environment becomes less suitable for AIS, their management will be positively affected, and in areas experiencing new invasions, management will be negatively affected in terms of impact and expense. These species responses illustrate the need for monitoring and the sharing of monitoring data in coordinated information systems nationally. While the complexities and uncertainties associated with climate-change effects on AIS underscore the need for monitoring, coordinating information resources, and engaging in further research, state agencies can take some actions now to adapt AIS management to this additional challenge using existing information.

2.5.1. Zebra Mussels

The zebra mussel (*Dreissena polymorpha*) population has expanded from its point of introduction in the Great Lakes in 1988 to its current range that includes 473 lakes, the five Great Lakes, and numerous rivers in 23 states, and most recently, it has been found in aquatic ecosystems in Nevada (Benson and Raikow, 2007). Zebra mussels form dense aggregates on hard substrates, altering invaded ecosystems by consuming native phytoplankton and other species in the water column and significantly reducing biomass. This adversely affects the consumed species and also alters food web patterns and changes water properties by increasing water clarity and light penetration. Often zebra mussels settle in water supply pipes of industrial and agricultural facilities, constricting flow and damaging equipment. Taken together, the zebra mussel and the quagga mussel (*Dreissena bugensis*) (another Great Lakes invader that causes similar impacts and that has a range that is expanding) are estimated to cause $1 billion in damages and costs annually (Pimentel, 2003).

Currently, there have been almost no successful mechanisms to selectively eradicate zebra mussels once a population has been established in a water body.[1] Therefore, prevention is the key tool to decreasing zebra mussel invasions. Zebra mussels spread by passive transport, in ballast and bilge water, and by attachment to boat hulls and other equipment. Important

---

[1]The Virginia Department of Game and Inland Fisheries eradicated zebra mussels from Millbrook Quarry by injecting twice the amount of potassium needed to kill zebra mussels over a three week period in the winter of 2006, four years after the first report of zebra mussels was submitted to the agency. For more information see: Virginia Department of Game and Inland Fisheries. Millbrook Quarry zebra mussel eradication. Available online at http://www.dgif.state.va.us/zebramussels/[accessed June 6, 2007].

prevention measures include inspecting and washing boats and dumping live bait and bilge water onto land. Because of the possibility of spread by recreational boaters and anglers, education and outreach are also important prevention tools. The 100th Meridian Initiative is one example of an interstate cooperative program that educates the public to prevent zebra mussel spread. This organization posts signs and brochures along highways and at boat ramps to teach the importance of cleaning and inspecting boats. Several states also have boat inspection sites, put out news releases, give presentations, educate divers, and train port of entry personnel. Missouri uses a variety of measures, such as installing Traveler Information Stations to advise boaters to clean their boats and working with bait shops to spread their message.

Freshwater lakes and streams in the northern U.S. may be available for recreation for longer periods of the year because of increasing temperatures. This would, in turn, extend the period of time during which recreational boaters and anglers could disperse zebra mussels in the northern U.S. Educational efforts such as those undertaken by the 100th Meridian Initiative may increase in importance, especially in the northern U.S. regions that are not yet infested with zebra mussels.

While higher latitudes and altitudes in the U.S. and Canada may become more suitable for zebra mussel invasion, habitats at the southern extent of its range may become less suitable. As temperatures rise, so do metabolic rates in zebra mussels. Unlike some species, zebra mussels have little capacity for metabolic adjustment to temperature change (Alexander et al., 1994). As turbidity increases, zebra mussel oxygen consumption drops, which may be due to increased undigestible particles clogging gills. Based on these results, Alexander et al. (1994) hypothesized that the most stressful conditions for zebra mussels would be high temperature and high turbidity conditions. Climate change may lead to these high temperature conditions in low altitude and latitude rivers and lakes in the U.S., making these habitats less suitable for zebra mussels; if these changes are combined with increased turbidity from altered precipitation and/or land use patterns, conditions may become too stressful for zebra mussels in these habitats. In addition, disturbances that result in die off of adults and decreased recruitment of 1-year old juveniles have been shown to stabilize zebra mussel populations (Strayer and Malcom, 2006). If changes in hydrology due to climate change include more intense flooding, this type of population stabilization that limits population size may occur more frequently, versus a more cyclic dynamic that can include very high densities. Management of more stable populations may be easier and impacts also may be more stable (Strayer and Malcom, 2006).

2.5.2. Water Hyacinth

Water hyacinth (*Eichhornia crassipes*) is a tropical aquatic plant native to Brazil that has invaded many countries (Charudattan, 2001). Considered one of the most problematic weeds in the world, it is highly invasive in southern states, Hawaii, and California (Ramey, 2001). As a floating weed, water hyacinth grows quickly, faster than any other saltwater, freshwater, or terrestrial vascular aquatic plant (Masifwa et al., 2001; Toft et al., 2003). Forming thick mats, water hyacinth rapidly over takes water bodies and significantly blocks water ways (Pimentel et al., 2000; Charudattan, 2001).

Control efforts are intensive and expensive. Florida spends $15 million a year to control three aquatic invasive plants including water hyacinth (Charudattan, 2001). Florida also mandated a coordinated control effort for water hyacinth including biochemical and chemical control measures and surveys, which has been very successful at controlling water hyacinth. Biocontrol methods involving weevil species also have proven successful in other parts of the world including Lake Victoria, Africa. Prevention, early detection, and regional coordination are critical for preventing aquatic weed invasions including water hyacinth (Charudattan, 2001).

Though water hyacinth is relatively cold tolerant and can survive in open waters (Charudattan, 2001), it cannot withstand winter temperatures in more northern states (Ramey, 2001). Climate-change impacts may enable both the spread and establishment of water hyacinth within states as well as into more northern states. Increased rainfall and hurricane intensity could result in more frequent and intense flooding events, which can facilitate its dispersal (Michener et al., 1997). Water hyacinth are able to survive these types of extreme events and can reestablish and colonize both in up- and down-stream systems (Center and Spencer, 1981). The increased frequency and intensity of disturbance events may create unsuitable conditions for native species, making ecosystems even more vulnerable to invasion by water hyacinth and enabling its spread. Water hyacinth is already present as an annual colonizer in some northern states, and warmer waters may enable and encourage its spread and establishment further north. Of particular concern are nurseries in northern states that sell water hyacinth for water gardens; plant escapes are a common mechanism of spread (Charudattan, 2001). These nurseries may become a viable pathway for water hyacinth as conditions in northern states become more suitable for water hyacinth survival.

2.5.3. Common Reed

Phragmites australis, the common reed, is prevalent on the Atlantic Coast and is rapidly spreading westward and northward. It is native to some regions of the U.S., but the invasive strain is believed to have been introduced from Europe in the late 1800s (Blossey et al., 2002). It is most abundant on the Atlantic coast and is expanding in the Midwest. Wilcox et al. (2003) mapped changes in *Phragmites* coverage between 1945 and 1999 by selecting nine different years to examine using aerial photos from the Great Lakes region. GIS maps show that its distribution expanded and contracted over that time period, but that it increased exponentially from 1995 to 1999. Wilcox et al. (2003) hypothesize that expansion will continue quickly through the Great Lakes. In addition, Bertness et al. (2002) demonstrate that increased nitrogen from shoreline development is facilitating expansion of *Phragmites* across New England salt marshes.

Phragmites control activities are important for wetland restoration projects. Methods to control *Phragmites* include biocontrol, flooding, non-specific herbicide control, cutting, and/or burning (Ailstock et al., 2001). Most states carry out herbicide applications in conjunction with other management techniques, such as mechanical removal, burning, or induced tidal flooding. Ohio, Delaware, and Virginia have had success applying herbicides aerially, and other states are considering this method. Several states carry out herbicide control measures on private lands through cost-sharing programs or through financial and technical assistance. Virginia has mapped *Phragmites* distribution within the state and uses this information to prioritize control and management actions.

Climate change may affect *Phragmites* control. *Phragmites* can tolerate brackish but not saline water (Asaeda et al., 2003), and therefore, sea level rise may help control this species and increase restoration success of some coastal wetlands. Areas predicted to be inundated by saltwater and to experience increased frequency of saltwater intrusion due to climate change may not be priority target areas for control actions. However, *Phragmites* populations also increase with higher-than-average ambient air temperatures (Wilcox et al., 2003), and thus other wetland areas may need to increase their control activities.

2.5.4. Green Crab

Invasive species have been introduced into marine ecosystems via a variety of vectors: ballast water and other shipping vectors; pet, aquaculture, and aquarium releases/escapes; opening of seaway canals between water bodies; and, to a lesser extent, research activities (Fofonoff et al., 2003; Siguan, 2003). The European green crab (*Carcinus maenus*) is a well-known invasive predator on both the western and eastern U.S. coasts and in various coastal regions of South Africa and Australia (Grosholz and Ruiz, 1996). The crustacean damages coastal fisheries by consuming juvenile native bivalves (Glude, 1955; Walton et al., 2002; Floyd and Williams, 2004). It is difficult to control and manage green crab populations. Lafferty and Kuris (1996) recommend developing and implementing biocontrol methods, although a biocontrol agent is currently not available (ANSTF Green Crab Control Committee, 2002). Maine has attempted physical control through fencing and manual control such as selective harvests, but the state has not been successful at controlling the green crab. In Maine, winters with below average temperatures have been the only thing that has succeeded in diminishing green crab populations (Dominie Consulting and Maine Interagency Task Force on Invasive Aquatic Plants and Nuisance Species Technical Subcommittee, 2002).

Warmer water temperatures due to climate change may thus cause further expansion and establishment of green crabs in areas where they previously did not survive through the winter. In addition, green crabs may spread and establish more easily in new areas as changes in climate reduce native bivalve populations. Climate-change effects also may exacerbate the impacts of green crabs to cause further habitat degradation. For example, native populations of commercially important clams (e.g., *Mya arenaria*, *Nutricola tantilla*, *Nutricola confusa*), already stressed by green crabs and other invasive species, may not be able to withstand the added stresses of climate-change effects, which include temperature changes, altered precipitation regimes, and altered patterns of wind and water circulation (Kennedy et al., 2002).

3. INFORMATION AND RESEARCH NEEDS AND GAPS

A comparison of information available in the scientific literature with the needs of decision-makers and state invasive-species managers reveals that there is a significant need for scientific, multi-factor, long-term studies to more fully understand the interactions between climate change and invasive species. Also, some specific data needs could be addressed quickly to enable managers to adapt AIS management practices for particular species and ecosystems in a changing climate. Section 3.1 outlines the immediate information needs of invasive-species managers to begin addressing the effects of climate change. Section 3.2 discusses scientific research needed to develop a more comprehensive understanding of the interactions between climate and invasive species under changing conditions.

Information needs for managers were determined based on three sources: (1) a synthesis of comments from an U.S. EPA-sponsored meeting attended by researchers and state invasive-species managers in June 2006, entitled "Assessing Gaps and Needs for Invasive Species Management in a Changing Climate," (2) our 50-state inventory of AIS management programs (see Appendix A: Aquatic Invasive Species Programs and Activities) and activities, and (3) a review of state and regional AIS management plans (see Appendix B, State Aquatic Invasive Species Management Plan Summaries and Appendix C, Regional Aquatic Invasive Species Management Plan Summaries). Research needs on invasive species and climate change interactions also are drawn from the workshop and the 50-state inventory as well as from a review of scientific literature on the effects of climate change on invasive species.

3.1. INFORMATION NEEDS FOR STATE MANAGERS

This section summarizes the information needed by state AIS managers to adapt management practices under changing conditions. Information needs are discussed according to several corresponding chapters in the National Invasive Species Management Plan: Leadership and Coordination, Prevention, EDRR, Control and Management, Restoration, and Information Management.

3.1.1. Information Needs for Effective Leadership and Coordination in a Changing Climate

Leadership and coordination within and among states and regions on invasive-species and climate-change issues are essential both for improving effectiveness of management efforts and for increasing awareness and understanding of these issues more generally. The need for better communication among states is a common concern among managers. For example, managers in Georgia have identified a need for interstate communication to prevent people from traveling across borders with illegal invasive species. This type of communication will be even more important as conditions change. Sharing information such as monitoring data among and within states and regions will help improve prevention and early detection efforts. Invasive species councils will be crucial to this effort; however, additional mechanisms and institutions would facilitate leadership and coordination on climate change and AIS issues at both state and regional levels.

Information and research needs for leadership and coordination under a changing climate may include the following:

- Identify AIS and climate-change leaders in each state to promote the importance of considering AIS and climate change together.

- Understand how other states are already cooperating on climate-change or invasive-species issues by examining existing channels (e.g., invasive-species councils) to share information on AIS and climate change and other mechanisms to facilitate the transfer of information (e.g., regular meetings, workshops, distribution lists, databases).

- Identify which structures, institutions, and/or policies work best across agencies and allow flexibility under changing conditions (e.g., flexibility in numbers or types of people working on issues and flexibility within legal authorities).

- Understand the consistencies and inconsistencies among states' laws that could affect the ability of states' agencies to cooperate both within and among states (e.g., problems and solutions affecting multiple media managed by different divisions or agencies).

- Prioritize invasive-species issues and concerns, in light of changing conditions.

- Identify existing, applicable adaptive management strategies that may guide state efforts to begin addressing climate change considerations in AIS management.

3.1.2. Information Needs for Effective Prevention Activities in a Changing Climate

Effective prevention methods are fundamental to stemming the tide of AIS. Prevention strategies will need to be adapted based on predicted and observed climate-change impacts. Thus, managers will need climate information as it relates to pathways, prediction and risk analyses, and monitoring.

3.1.2.1. *Information Needs Related to Pathways*

States need information on how the effects of climate change (e.g., changes in precipitation patterns and temperature) interact with vectors and pathways of AIS transport. Massachusetts, for example, is particularly concerned about aquatic plants sold by nurseries that could escape and become established as water temperatures increase. An understanding of this interaction will help state AIS managers prioritize monitoring, inspection, education, and regulatory efforts. However, identifying AIS pathways can be challenging, especially in light of anticipated climate changes. More information is needed to understand how vectors and pathways are influenced by climate change, and this information needs to be shared among states.

Information needs for pathways and vectors under a changing climate may include the following:

- Identify current priority pathways at state and regional levels and use existing environmental and biological data to identify how these pathways/vectors may change as temperatures and/or precipitation patterns change.

- Identify new pathways that will emerge under a range of potential climate-change conditions, including increased water temperatures, changes in precipitation, sea level rise, and changes in sea-ice cover. The factors to consider will vary depending on state and region.

- Identify species that will become invasive as conditions change in order to help target pathway analyses. Use data and information from other states with similar habitats and ecosystem types to extrapolate potential new invaders.

  o For example, extended warm temperatures in some areas due to climate change may result in an increase in recreational fishing, which could lead to a rise in boat traffic (an important AIS vector). Understanding the AIS implications of the emergence of these pathways, such as an increase in water hyacinth or zebra mussel introductions, and behavioral responses will be important information for managers adapting prevention and monitoring strategies.

- Determine how pathway/vector analyses can be modified to account for climate-change effects (e.g., temperature, precipitation, and sea level changes) and provide accurate predictions.

- Incorporate climate change information into models and systems that predict changes in pathways and transfer mechanisms.

3.1.2.2. *Information Needs Related to Prediction Models and Risk Analyses*

Besides understanding the interaction between climate change effects and vectors, managers need to know how species and habitats will respond to climate change (e.g., range expansion, ability for species to establish, habitat vulnerability to invasion). By integrating biological and ecological data on AIS with different climate-change scenarios, such as warmer summer temperatures, altered precipitation regimes, earlier snowmelt, or increased carbon dioxide levels, scientists and managers can improve risk analyses and prediction models.

Information needs for prediction models and risk analyses under a changing climate may include the following:

- Determine how existing invasive-species prediction models may be modified to incorporate climate-change data (e.g., water temperature, timing of precipitation, dissolved oxygen content, and sea level rise). Specifically, it will be important to integrate known data about the biology of AIS into mechanistic, spatially-explicit models that include relevant climate-change parameters. Temporal and spatial scales of AIS spread and establishment will also need to be considered.

- Develop new models to improve predictions of species responses to climate changes in order to provide managers with some expectations for ecosystem changes. Consider habitat alterations caused by climate change, especially thresholds in aquatic habitats, and the interactions between species' adaptive capacities, their shifting climatic boundaries, and the shifting landscape that will lead to new potential distributions.

- Establish baseline datasets with information on existing AIS at state and regional levels in order to allow quantitative statistical analysis across climate-change scenarios.

- Identify AIS not yet found in northern climates that have temperature tolerances that would allow them to overwinter as northern climates become milder (i.e., there is a need for information on temperature tolerances of species and on how these tolerances may change over time).

- Research how climate change may affect the conditions that may lead to invasion (e.g., disturbed habitat, decreased native biodiversity, and altered light availability). Research will need to focus on both species and habitat characteristics.

  o For example, two questions to consider are, will increased intense weather events (e.g., hurricanes, floods) lead to an increase in disturbed habitats that could facilitate invasion by AIS? and will certain AIS be more prone to invade these habitats than other AIS?

- Identify mechanisms to integrate climate change parameters (e.g., water temperature, dissolved oxygen content, sea level rise) into risk analyses to more accurately determine the threat of a species establishment and spread within an area.

- Assess the risk that non-native species currently allowed into the U.S. may become invasive and/or expand their ranges in response to climate change. Coordination with other states and other types agencies will be important in addressing this need, because decisions about which species to allow are not always made by the same agencies that monitor and manage AIS.

3.1.2.3. *Information Needs Related to Monitoring*

Monitoring efforts will need to be adapted to ensure effective identification of potential new AIS, as well as existing AIS present at low levels. Collaborating with neighboring states to share monitoring data may facilitate the process of identifying potential new AIS.

Information needs for monitoring under a changing climate may include the following:

- Develop, establish, and fund strategically placed and comprehensive monitoring systems.

    o Integrate or coordinate monitoring systems among states.

    o Design monitoring systems to incorporate the potential effects of climate change, especially temperature and precipitation changes that influence climatic boundaries of AIS.

    o Establish monitoring baselines to detect changes in both climate and AIS.

- Use research on encroaching species, climate-change effects on ecosystems, and new pathways that may emerge as a result of climate change to determine priority pathways, areas, and species to monitor.

    o For example, if pathway monitoring efforts in a state focus primarily on aquatic plant imports, but recreational boating and fishing are expected to increase as temperatures stay warmer for longer periods, then monitoring efforts and techniques may need to be developed that focus on boat inspections and bait usage.

- Use information on how habitats and ecosystems will respond to climate change (i.e., become more vulnerable to invasion) to help identify priority areas for monitoring.

- Use information on how species ranges and distributions will respond to climate change (i.e., expanding ranges) to help identify priority areas for monitoring.

- Modify monitoring methods to identify effects from climate change (e.g., temperature, precipitation, and sea level changes) and possibly distinguish between climate variability (e.g., drought cycles) and long-term climate change.

- Develop a core set of indicators for state managers to use when monitoring for AIS under changing conditions.

3.1.3. Information Needs for Effective Early Detection and Rapid Response in a Changing Climate

Research to inform coordination and prevention also will help improve EDRR efforts under climate change conditions. However, additional information needs specific to EDRR also exist:

- Evaluate existing state EDRR capabilities (i.e., quarantine authority, emergency powers, and border control capacity) in order to determine effectiveness in addressing invasive threats resulting from changing conditions.

- Develop an effective EDRR system (if existing system is insufficient) that anticipates barriers and deals with them before any new species arrives, so response can be swift and effective. The system will need to include successful mechanisms for inspections and response. An EDRR system designed in this way will allow state agencies to detect potential invaders that may be more prevalent as conditions change.

- Collect information on altered species ranges and/or pathways under climate change to help identify where to target early detection monitoring efforts.

- Ensure priority lists of AIS are updated regularly to reflect changes in species as conditions change.

- Develop rapid response protocols for species that are predicted to become more invasive under a changing climate.

3.1.4. Information Needs for Effective Control and Management in a Changing Climate

Control and management practices also will need to account for climate change to ensure effective and successful control and eradication of AIS. There is already growing recognition by state managers of the need for more research on control methods and technologies for a wide range of species, such as zebra mussels, Eurasian water milfoil, *Phragmites,* apple snails, etc. Thus, as a part of the process to identify appropriate control techniques for specific species, scientists and managers also should study how climate change may impact these control methods.

Information needs for control and management under a changing climate may include the following:

- Research the performance of biological, chemical, and mechanical controls under various climatic conditions (e.g., increased temperatures, hydrology changes, and altered water chemistry).

- Determine which biological or chemical control methods will be most adaptable, or will remain robust, under climatic variability and change.

- Identify existing mechanical controls that adequately consider climate change.

- Develop guidelines on how climate change may affect different biocontrol species.

3.1.5. Information Needs for Effective Restoration in a Changing Climate

Managers also will need climate information to ensure restoration plans are adequately designed to re-establish ecosystem processes and be successful over the long-term. Information needs for restoration under a changing climate may include the following:

- Research how best to restore ecosystem processes in invaded areas, such as sediment and nutrient transport and how restoration of these processes could be affected by climate change (e.g., how salinity, nutrient, and hydrological regime changes may impact the system's nutrient transport capabilities).

- Determine which species used in restoration projects will remain viable under future climatic conditions.

- Conduct studies to understand the types of feedbacks that may exist between climate change factors and invasibility so that restoration plans can adequately account for climate change conditions. For example, coastal marsh restoration is dependent on water levels. With sea level rise, marsh restoration projects could be destroyed.

3.1.6. Information Needs for Effective Information Management in a Changing Climate

An information system that documents, evaluates, and monitors AIS impacts is imperative to prevention, early detection, and control efforts. An information management system also must include distribution and establishment data, and correlate and synthesize data from many sources. Various national information systems for tracking and organizing data on invasive species exist, including the Nonindigenous Species Database Network (NISbase) and the Nonindigenous Aquatic Species (NAS) information resource. NISbase is working to bring fragmented information on invasive species together into a single database that can be queried. The NAS system focuses on AIS and will work with states to make sure their specific needs are met. In moving forward with these systems, coordination states and other databases will be important to ensure that the information systems meet the needs of all users, especially as climatic conditions change. Climate-change data (e.g., water temperature, salinity levels, and other hydrological parameters) will need to be included to make systems more robust and accurate. Systems will need to be made dynamic and updatable to reflect changes in species distributions and establishment that may be caused by climate change (Lee et al., 2008).

3.1.7. Information Needs for Effective Public Education in a Changing Climate

Public education activities will need to include information on climate change and its likely effects on aquatic ecosystems and AIS. These activities could also be used to highlight how states are preparing to deal with these effects and what additional actions may be needed.

3.2. RESEARCH NEEDS ON AQUATIC INVASIVE SPECIES AND CLIMATE CHANGE

In Section 3.1 above, we discussed the immediate information and data needs of conservation managers to begin addressing climate-change conditions. Below we address broader scientific research needed to develop a more comprehensive understanding of the interactions between climate and invasive species under changing conditions. These research needs are derived from the literature review conducted to develop this report and also a synthesis of the June 2006 workshop, "Assessing Gaps and Needs for Invasive Species Management in a Changing Climate." The workshop informed both this section on research needs, which are directed towards the scientific community and represent more basic science needs, and the previous section (3.1) on information needs for managers, which are more immediate needs representing more applied scientific questions. The research needs in this section (3.2) are necessarily broad in scope, demonstrating the paucity of information on climate-change and invasive-species interactions. For all research needs, however, climate-change data will be most useful when it is tied to specific regions, and thus, to AIS that occur in those areas. For example, research on the impacts of climate change in western North America projects that earlier snowmelt due to increasing temperatures will impact stream flow (Stewart et al., 2004), an impact that will be important when identifying how AIS may respond to climate change in that region. However, regional climate-change modeling and smaller-scale projections of effects on specific watersheds are at the current edge of scientific research; therefore, more detailed assessments of effects on specific AIS in specific places is not yet possible.

3.2.1. Climate-Change Effects on Invasive Species

Research is needed on the effects of climate change on invasive species in all aspects of the invasion pathway as discussed below.

Pathways and Vectors

- Effects of climate change (e.g., water and air temperature changes, precipitation patterns, and sea level rise) on AIS pathways and vectors, including new pathways and changes in existing pathways.

Establishment and Spread

- Ecosystem feedbacks between climate change and conditions favorable to AIS establishment and spread.

- Effects of climate change on current high priority invasive species, both positive and negative, in terms of changing distributions and impacts.

- Effects of carbon dioxide on freshwater ecosystems and AIS.

- Effects of changing precipitation patterns, such as flood and drought frequencies, on AIS establishment, habitat availability, and spread.

- Effects of increasing temperature on AIS establishment, habitat availability, and spread.

Ecosystem Susceptibility
- Climate-change effects on the susceptibility of aquatic ecosystems to invasion by AIS.

  o For example, conduct studies to clarify the complex interactions among factors affecting species distribution and to determine whether climate change will increase susceptibility of habitats and regions to invasions, including assessment of positive interactions among non-native species and circumstances under which biodiversity may provide a barrier to invasions;

- Restoration and resilience effects on the susceptibility of ecosystems to invasion by AIS in the face of climate change.

- Studies of ecosystems recovering from disturbed states to understand the impacts of AIS on native species under changing climatic conditions.

- Climate-change effects on different types of coastal and ocean currents and resulting effects on the spread and distribution of AIS and their impacts to coastal ecosystems.

3.2.2. Interacting Stressors

Additional research is needed on how other stressors (e.g., land-use change, overfishing, and pollution) interact with climate change and on how these interactions affect invasive species, including

- How other stressors (e.g., land-use change, overfishing, and pollution), climate change, and susceptibility to invasion are related.

- How increasing temperatures, water quality problems resulting from pollution, and AIS may interact and the feedbacks that may occur among these factors.

- How changing precipitation patterns, water quality problems resulting from pollution, and AIS may interact and the feedbacks that may occur among these factors.

- How interactions between climate and land-use change may affect distribution, spread, establishment, and impacts of AIS.

- How development patterns may change under climate change and resulting effects on AIS.

3-9

- How climate and overfishing impacts interact to affect AIS.

- How other factors (e.g., poor water quality) may facilitate the establishment and spread of AIS under climate change.

Mack et al. (2000) note that research on just invasive species dates back only a few decades, and more research is needed particularly on the epidemiology of invasive species so that predictions may be more accurate. This information will also be important for understanding how invasive species may respond to changing conditions. Dukes and Mooney (1999) identify a need to study climate-change effects on invasive-species distribution, while Byers (2002) suggests studying the impacts of non-native species on native species as the system recovers from a disturbed state to more natural conditions. Overall, researchers conclude that we still need to conduct a significant amount of research on invasive species and climate change in order to address many of the information needs of managers.

4. CONCLUSIONS, RECOMMENDATIONS, AND NEXT STEPS

Both invasive species and climate change are major ecosystem stressors. Although not well understood, particularly in aquatic ecosystems, the interaction of these stressors may exacerbate the effects of each. In order to design and conduct effective AIS management, state managers should put in motion efforts that will allow them to consider the projected effects of climate change on AIS prevention, control, and eradication actions. This assessment of the current status of climate in AIS management underscores the need to consider climate-change effects in every part of AIS management plans and programs in order to address AIS threats effectively.

Incorporation of climate-change information is important for every state program with AIS responsibility. Indeed, adapting AIS management practices will allow states to better prevent and control AIS invasions under changing conditions, as well as maximize the effectiveness and efficiency of each state dollar spent on AIS management activities. However, our review shows that, with few exceptions, states have not put in place adaptive management strategies that incorporate climate-change information. This result is not surprising since there currently is no legislative mandate to consider climate change in these activities. Our review does highlight, however, that there is considerable capacity to adapt existing plans to include climate-change information. For example, most plans include provisions for revising and updating the plan allowing new information to be incorporated, funding specific management goals and activities, and implementing or modifying existing monitoring activities.

Despite the lack of a legislative mandate, several states have taken or are taking steps to consider climate change (e.g., Alaska, Arkansas, Kansas, Maine, and Wisconsin) (see Appendix A, Aquatic Invasive Species Programs and Activities). For example, Arkansas is concerned about the potential for water hyacinth to overwinter as water temperatures warm (see Appendix A, Aquatic Invasive Species Programs and Activities), and the Arkansas Plant Board recently added water hyacinth to the state's noxious weed and prohibited plant lists (Arkansas State Plant Board, 2007). Taking this step now to prevent water hyacinth introductions could help prevent its future spread due to warmer conditions.

For the majority of states not addressing climate change, a significant factor may be the lack of reliable science-based information to inform AIS managers and decision-makers in designing and implementing their plans, programs, and activities. Fortunately, many state plans include research tasks that incorporate at least some capacity to examine changing conditions, thus potentially reinforcing a hope for more information. In addition, the structure and substance of some state plans suggest that managers are thinking about environmental change and may be interested in including climate change more explicitly if enabled to do so cost-effectively.

Scientific research, development of models and predictors, and data collection should be conducted in order to provide managers with the tools and information they need to conduct effective prevention, control, and eradication of AIS. Information needs include both immediate data needs and long-term research to better understand the complex interactions between climate change and aquatic invasions.

Below we summarize five recommendations that are designed to maintain and improve state AIS management programs and activities in a changing climate. For each recommendation, it will be important for states to consider how to collect data, test hypotheses, and record results; how to change their management plans; and how to coordinate action to move forward on each recommendation. Using an adaptive-management framework will make these recommendations easier to implement as states address climate change.

4.1. INCORPORATING CLIMATE INTO AQUATIC INVASIVE SPECIES LEADERSHIP AND COORDINATION ACTIVITIES

Invasive-species councils, or lead state agencies in the absence of councils, can incorporate climate considerations into their management plans. This might be initiated by conducting facilitated meetings and/or workshops to identify specific management strategies and research needs to inform management strategies. State councils also could work with one another to share information on climate-related data across regions. Coordination and information sharing among states will also facilitate the implementation of activities that are adapted to climate change effects. State and federal agencies also could collaborate in areas such as AIS data collection, specifically where the spatial scale of the biological and environmental data needed by the federal government may be more efficiently collected by a state. In turn, the data provided to the federal government by states may be used in modeling scenarios that also would benefit state AIS management efforts.

4.2. IDENTIFYING AQUATIC INVASIVE SPECIES THREATS UNDER CHANGING CONDITIONS

In order to effectively prevent invasions that might result from or be influenced by climate-change factors, a first step should be to identify specific AIS threats, including new pathways and vectors, which may result as environmental conditions such as water and air temperatures, precipitation patterns, or sea levels change. In implementing this step, the initial focus should be placed on state priority AIS. Coordination among states to share information on species and pathways will aid in data collection and implementation of prevention activities. State collaboration could be carried out through regular meetings between invasive-species councils and/or agency AIS personnel.

Comprehensive monitoring systems that can detect new AIS, new impacts, and range changes as a result of climate change must be developed, established, and funded. Ideally, these systems should be accessible to managers within a state and among states to ensure dissemination of important information. The systems should also be easy to update as more information on AIS and climate change becomes available. Pathway analysis and species prediction models also need to be modified and/or developed to address the multiple factors that drive invasions. Models that incorporate predictions of changes in air temperature, water temperature, precipitation, and sea level, may provide highly useful projections on changes in the movement of species' range boundaries. In response, regulatory requirements and education efforts can be adjusted accordingly. Each of these steps could benefit from additional research that specifically addresses how current practices may need to change in light of climate change.

4.3. IDENTIFYING VULNERABLE ECOSYSTEMS UNDER CHANGING CONDITIONS AND DESIGNING RESILIENT RESTORATION

Effective AIS prevention efforts must include identification of ecosystems that may be more vulnerable to invasion under changing environmental conditions. This effort should be complemented by identification of key restoration opportunities. Restoration of ecosystems is an important aspect to comprehensive prevention strategies, as robust habitats are less vulnerable to invasion. For these reasons, restoration should be designed to thrive under, or at least withstand, the changing temperatures, precipitation patterns, and sea level changes that are predicted to result from climate change. Both identifying vulnerable ecosystems and restoring ecosystems to become less vulnerable are activities that would benefit from additional research that includes climate change interactions. As a first step, states could begin collecting data on vulnerable ecosystems and restoration techniques appropriate for these ecosystems; if the data already exist, these areas could be used to test restoration techniques in an adaptive-management framework.

4.4. IMPROVING CONTROL MEASURES UNDER CHANGING CONDITIONS

States should evaluate control measures for efficacy under the altered conditions that may result from a changing climate and should adjust AIS management priorities and plans accordingly. Biological, chemical, manual, and mechanical control methods may all be affected by climate change. Managers will need to coordinate with scientists to obtain any existing information on different control methods and how climate-change effects, such as increased temperatures and altered precipitation patterns, may affect their efficacy, so that states may be better prepared to adapt their control programs. Because more research may be needed to identify how climate change will affect control mechanisms, state managers will need to coordinate with scientists to ensure that research is focused on the most critical and vulnerable control methods.

4.5. MANAGING INFORMATION UNDER CHANGING CONDITIONS

States designing AIS information management systems should include the capacity to account for changing conditions by collecting and tracking climate change data (e.g., water temperature, salinity levels, and water chemistry). Including this information will ensure robustness and accuracy of information management systems under changing conditions. States also should check existing information systems such as NISbase and the NAS information resource to prevent duplicative efforts and also to benefit from information already collected. States' managers also should take steps to ensure that the data collected by states is integrated into these existing systems.

4.6. NEXT STEPS FOR RESEARCHERS AND MANAGERS

Although there is much to be done for states and their partners to begin to address climate change in AIS management, the importance of making a concerted movement is underscored by the findings of this report. State AIS managers have concrete needs for information and data; the research community, including universities, government agencies, nongovernmental organizations, and private groups, has capability to address these needs. But specific financial and programmatic support for all of these activities does not now exist. However, even under the current circumstances, states have some significant options for incorporating climate considerations to a greater extent into their current AIS efforts.

An adaptive-management framework may be the most appropriate framework for states to use to begin incorporating climate-change information into management plans and programs. Adaptive management involves testing the effectiveness of different management methods. This testing will be important because of the high levels of uncertainty about specific temporal and spatial effects of climate change. In addition, coordinating research with state needs and activities will ensure that the design and implementation of an adaptive-management framework is effective at addressing and anticipating climate-change effects (Bierwagen et al., 2008).

An additional, important step for states is coordination among state and regional invasive-species councils and state agency personnel that manage AIS. This collaboration will facilitate information sharing on various management activities that will likely be affected by climate change, including pathway identification, monitoring data, and control mechanisms. In addition, agency staff and AIS coordinators would receive valuable information from reviewing current prevention, control, and eradication activities, as well as planned action items, for their potential vulnerability to climate change; identifying specific data and information needs; and modifying current strategies where feasible and when climate information is available from the growing body of scientific literature or from knowledgeable practitioners and researchers.

REFERENCES

Aerts, R; Cornelissen, JHC; Dorrepaal, E. (2006) Plant performance in a warmer world: general responses of plants from cold, northern biomes and the importance of winter and spring events. Plant Ecol 182(1–2):65–77.

Ailstock, MS; Norman, CM; Bushmann, PJ. (2001) Common reeds *Pragmites australis*: control and effects upon biodiversity in freshwater nontidal wetlands. Restor Ecol 9(1):49–59.

Alexander, JE, Jr; Thorp, JH; Fell, RD. (1994) Turbidity and temperature effects on oxygen consumption in the zebra mussel (*Dreissena polymorpha*). Can J Fish Aquat Sci 51:179–184.

Asaeda, T; Manatunge, J; Fujino, T; Sovira, D. (2003) Effects of salinity and cutting on the development of *Phragmites australis*. Wetlands Ecol Manage 11(3):127–140.

ASPB (Arkansas State Plant Board). (2007) Regulations on plant diseases and pests. Circular 11. ASPB, Little Rock, Arkansas. Available online at http://www.plantboard.org/plant_pdfs/CIRCULAR%2011_EMERGENCY%20RULES_%20CHENIERE_SEED%20TESTING_RICE%20REGS_%20MB_TURF_NURSRY_DEC28_2006.pdf.

Barrett, SCH. (2000) Microevolutionary influences of global changes on plant invasions. In: Mooney, HA; Hobbs, RJ; eds. Invasive species in a changing world. Washington, DC: Island Press; pp. 115–139.

Barry JP; Baxter, CH; Sagarin, RD; et al. (1995) Climate-related, long-term faunal changes in a California rocky intertidal community. Science 267(5198):672–675.

Benson, AJ; Raikow, D. (2007) *Dreissena polymorpha*: USGS nonindigenous aquatic species database (last modified January 18, 2008), Gainesville, FL. Available online at http://nas.er.usgs.gov/queries/FactSheet.asp?speciesID=5.

Bertness, MD; Ewanchuk, PJ; Silliman, BR. (2002) Anthropogenic modification of New England salt marsh landscapes. Proc Natl Acad Sci USA 99(3):1395–1398.

Bierwagen, B; Thomas, R; Kane, A. (2008) Aquatic invasive species management plans: an assessment of their adaptive capacity to integrate climate change. Conserv Biol: in press.

Blossey, B; Schwarzlander, M; Halfiger, P; et al. (2002) Common reed. In: Van Driesche R; Lyon, S; Blossey, B; et al.; eds. Biological control of invasive plants in the eastern United States. United States Department of Agriculture Forest Service Publication; FHTET-2002-04. Available online at http://www.invasive.org/eastern/biocontrol/9CommonReed.html.

Braby, CE; Somero, GN. (2006) Following the heart: temperature and salinity effects on heart rate in native and invasive species of blue mussels (genus *Mytilus*). J Exp Biol 209(Pt 13):2554–2566.

Bryant, SR; Thomas, CD; Bale, JS. (2002) The influence of thermal ecology on the distribution of three nymphalid butterflies. J Appl Ecol 39(1):43–55.

Buchan, LAJ; Padilla, DK. (1999) Estimating the probability of long distance overland dispersal of invading aquatic species. Ecol Appl 9(1):254–265.

Byers, JE. (2002) Impact of non-indigenous species on natives enhanced by anthropogenic alteration of selection regimes. Oikos 97(3):449–458.

Byers, JE; Noonburg, EG. (2003) Scale dependent effects of biotic resistance to biological invasion. Ecology 84(6):1428–1433.

Cao, L; Caldeira, K; Jain, AK. (2007) Effects of carbon dioxide and climate change on ocean acidification and carbonate mineral saturation. Geophys Res Lett 34:L05607, doi:10.1029/2006GL028605.

Carlton, JT. (2000) Global change and biological invasions in the oceans. In: Mooney, HA; Hobbs, RJ; eds. Invasive species in a changing world. Washington, DC: Island Press; pp. 31–53.

Carlton, JT; Ruiz, GM. (2005) Vector science and integrated vector management in bioinvasion ecology: conceptual frameworks. In: Mooney, HA; Mack, RN; McNeely, JA; et al; eds. Invasive alien species. Washington, DC: Island Press; pp. 36–58.

Carroll, AL; Taylor, SW; Régnière, J; et al. (2003) Effects of climate change on range expansion by the mountain pine beetle in British Columbia. In: Shore, TL; Brooks, JE; Stone, JE; eds. Mountain pine beetle symposium: challenges and solutions. Information Report BC-X-399. Victoria, British Columbia: Natural Resources Canada, Canadian Forest Service, Pacific Forestry Centre; p. 223–232. Available online at http://www.for.gov.bc.ca/hfd/library/MPB/carroll_2004_effects.pdf.

Center, TD; Spencer, NR. (1981) The phenology and growth of water hyacinth (*Eichhornia crassipes*) in a eutrophic north central Florida USA lake. Aquat Bot 10(1):1–32.

Charles, H; Dukes, JS. (2007) Impacts of invasive species on ecosystem services. In: Nentwig, W; ed. Biological invasions. Ecol Studies 193. New York: Springer; pp. 217–337.

Charudattan, R. (2001) Are we on top of aquatic weeds? Weed problems, control options, and challenges. In: Riches,CR; ed. 2001 BCPC symposium proceedings no. 77: the world's worst weeds. Brighton, United Kingdom: The British Crop Protection Council; pp 43–68. Available online at http://plantpath.ifas.ufl.edu/People/Faculty/Charudattan/BioControl/PDF/AquaticWeedsChapter.pdf.

Chen, DX; Coughenour, MB; Eberts, D; et al. (1994) Interactive effects of CO_2 enrichment and temperature on the growth of dioecious *Hydrilla verticilala*. Environ Exp Bot 34:345–353.

Clavero, M; García-Berthou, E. (2005) Invasive species are a leading cause of animal extinctions. Trends Ecol Evol 20(3):110.

Cohen, AN; Carlton, JT. (1998) Accelerating invasion rate in a highly invaded estuary. Science 279(5350):555–558.

Cox, GW. (1999) Alien species in North America and Hawaii: impacts on natural ecosystems. Washington, DC: Island Press.

Curnutt, JL. (2000) Host-area specific climatic-matching: similarity breeds exotics. Biol Conserv 94(3):341–351.

Davis, MA; Grime, JP; Thompson, K. (2000) Fluctuating resources in plant communities: a general theory of invasibility. J Ecol 88:528–534.

Denny, MW; Paine, RT. (1998) Celestial mechanics, sea-level changes, and intertidal ecology. Biol Bull 194(2):108–115.

Diaz-Almela, E; Marba, N; Duarte, CM. (2007) Consequences of Mediterranean warming events in seagrass (*Posidonia oceanica*) flowering records. Global Change Biol 13(1):224–235.

Didham, RK; Tylianakis, JM; Hutchison, MA; et al. (2005) Are invasive species the drivers of ecological change? Trends Ecol Evol 20(9):470–474.

Dominie Consulting; Maine Interagency Task Force on Invasive Aquatic Plants and Nuisance Species Technical Subcommittee. (2002) State of Maine action plan for managing aquatic invasive species. Plan adopted by the Land

and Water Resources Council and the Maine Interagency Task Force on Invasive Aquatic Plants and Nuisance Species. Available online at http://www.maine.gov/dep/blwq/topic/invasives/invplan02.pdf [accessed December 28, 2007].

Dukes, JS; Mooney, HA. (1999) Does global change increase the success of biological invaders? Trends Ecol Evol 14(4):135–139.

Dukes, JS; Mooney, HA. (2004) Disruption of ecosystem processes in western North America by invasive species. Rev Chil Hist Nat 77:411–437.

Ehrenfeld, JG. (2003) Effects of exotic plant invasions on soil nutrient cycling processes. Ecosystems 6(6):503–523.

Eiswerth, ME; Donaldson, SG; Johnson, WS. (2000) Potential environmental impacts and economic damages of Eurasian watermilfoil (*Myriophyllum spicatum*) in western Nevada and northeastern California. Weed Technol 14(3):511–518.

ELI (Environmental Law Institute). (2002) Defining an invasive species. In: Halting the invasion: state tools for invasive species management. Washington, DC: ELI; pp. 27–32.

Executive Presidential Order. (1999) Executive order 13112 of February 3, 1999: invasive species. Federal Register 64(25):6183–6186. Available online at http://frwebgate.access.gpo.gov/cgi-bin/getdoc.cgi?dbname=1999_register&docid=fr08fe99-168.pdf.

Feely, RA; Sabine, CL; Lee, K; et al. (2004) Impact of anthropogenic CO_2 on the $CaCO_3$ system in the oceans. Science 30(5682):362–366.

Findlay, S; Groffman, P; Dye, S. (2003) Effects of *Phragmites australis* removal on marsh nutrient cycling. Wetlands Ecol Manage 11(3):157–165.

Floyd, T; Williams, J. (2004) Impact of green crab (*Carcinus maenas* L.) predation on a population of soft-shell clams (*Mya arenaria* L.) in the southern Gulf of St. Lawrence. J of Shellfish Res 23(2):457–462.

Fofonoff, PW; Ruiz, GM; Steves, B; et al. (2003) In ships or on ships? Mechanisms of transfer and invasion for nonnative species to the coasts of North America. In: Ruiz, GM; Carlton, JT; eds. Invasive species: vectors and management strategies. Washington, DC: Island Press; pp. 152–182.

Glassner-Shwayder, KM. (2000) Briefing paper: Great Lakes nonindigenous invasive species. A product of the Great Lakes nonindigenous invasive species workshop; October 20–21, 1999; sponsored by the U.S. Environmental Protection Agency, Great Lakes National Program Office. Available online at http://www.glc.org/ans/pdf/briefpapercomplete.pdf.

Glude, JB. (1955) The affects of temperature and predators on the abundance of the soft-shell clam, *Mya arenaria*, in New England. Trans Am Fish Soc 84:13–26.

Gordon, D. (1998) Effects of invasive, non-indigenous plant species on ecosystem processes: lessons from Florida. Ecol Appl 8(4):975–989.

Graham, RW; Lundelius, EL; Graham, MA; et al. (1996) Spatial response of mammals to late quaternary environmental fluctuations. Science 272(5268):1601–1606.

Green Crab Control Committee. (2002) Management plan for the European Green Crab. Grosholz, ED; Ruiz, GM; eds. Submitted to the Aquatic Nuisance Species Task Force, Arlington, VA. Available online at http://www.anstaskforce.gov/GreenCrabManagementPlan.pdf.

Grosholz, ED. (1996) Contrasting rates of spread for introduced species in terrestrial and marine systems. Ecology 77(6):1680–1686.

Grosholz, ED; Ruiz, GM. (1996) Predicting the impact of introduced marine species: lessons from the multiple invasions of the European green crab (*Carcinus maenasi*). Biol Conserv 78(1–2):59–66.

Hansen, MJ; Clevenger, AP. (2005) The influence of disturbance and habitat on the presence of non-native plant species along transport corridors. Biol Conserv 125(2):249–259.

Harrison, IJ; Stiassny, MLJ. (1999) The quiet crisis: a preliminary listing of freshwater fishes of the world that are extinct or "missing in action". In: MacPhee, RDE; ed. Extinctions in near time: causes, contexts and consequences. New York: Kluwer Academic/Plenum Publishers; pp. 271–332.

Hellmann, JJ; Byers, JE; Bierwagen, BG; et al. (2008) Challenges and opportunities for invasive species research in a changing climate: six responses and associated management strategies. Conserv Biol: submitted.

Helmuth, B; Broitman BR; Blanchette CA; et al. (2006) Mosaic patterns of thermal stress in the rocky intertidal zone: implications for climate change. Ecol Monogr 76(4):461–479.

Hobbs, RJ. (2000) Land use changes and invasions. In: Mooney, HA; Hobbs, RJ; eds. Invasive species in a changing world. Washington, DC: Island Press; pp. 55–64.

Hughes, TP; Baird, AH; Bellwood, DR; et al. (2003). Climate change, human impacts, and the resilience of coral reefs. Science 301(5635):929–933.

IPCC (Intergovernmental Panel on Climate Change). (2001) Summary for policy makers. In: Watson, RT; et al, eds. Climate change 2001: synthesis report. A contribution of Working Groups I, II, and III to the Third Assessment Report of the Intergovernmental Panel on Climate Change. Cambridge, United Kingdom: Cambridge University Press; pp 1–34. Available online at http://www.grida.no/climate/ipcc_tar/vol4/english/pdf/spm.pdf.

IPCC (Intergovernmental Panel on Climate Change). (2007) Climate change 2007: the physical science basis: summary for policymakers. Contribution of Working Group I to the Fourth Assessment Report of the Intergovernmental Panel on Climate Change. Geneva, Switzerland: IPCC. Available online at http://www.aaas.org/news/press_room/climate_change/media/4th_spm2feb07.pdf.

Kearney, M; Porter, WP. (2004) Mapping the fundamental niche: physiology, climate, and the distribution of a nocturnal lizard. Ecology 85(11):3119–3131.

Keller, RP; Lodge, DM; Finnoff, DC. (2007) Risk assessment for invasive species produces net bioeconomic benefits. Proc Natl Acad Sci USA 104:203–207.

Kennedy, VS; Twilley, RR; Kleypas, JA; et al. (2002) Coastal and marine ecosystems and global climate change: potential effects on U.S. resources. Arlington, VA: Pew Center on Global Climate Change. Available from http://www.pewclimate.org/docUploads/marine_ecosystems.pdf [accessed May 2007].

Kluza, DM; McNyset, KM. (2005) Ecological niche modeling of aquatic invasive species. Aquat Invaders 16(1):1–7.

Kolar CS; Lodge DM. (2000) Freshwater nonindigenous species: interactions with other global changes. In: Mooney, HA; Hobbs, RJ; eds. Invasive species in a changing world. Washington, DC: Island Press; pp. 3–30.

Kolar, CS; Lodge, DM. (2001) Progress in invasion biology: predicting invaders. Trends Ecol Evol 16(4):199–204.

Lacoul, P; Freedman, B. (2006) Recent observation of a proliferation of *Ranunculus trichophyllus* Chaix. in high-altitude lakes of the Mount Everest region. Arct Antarct Alp Res 38(3):394–398.

Lafferty, KD; Kuris, AM. (1996) Biological control of marine pests. Ecology 77(7):1989–2000.

Lake, JC; Leishman, MR. (2004) Invasion success of exotic plants in natural ecosystems: the role of disturbance, plant attributes and freedom from herbivores. Biol Conserv 117(2):215–226.

Lee, H; Reusser, DA; Olden, JD; et al. (2008). Integrated monitoring and information systems for managing aquatic invasive species in a changing climate. Conserv Biol: in review.

Leung, B; Lodge, DM; Finnoff, D; et al. (2002) An ounce of prevention or a pound of cure: bioeconomic risk analysis of invasive species. Proc R Soc Lond, Ser B: Biol Sci 269(1608):2407–2413.

Lite, SJ; Stromberg, JC. (2005) Surface water and ground-water thresholds for maintaining *Populus–Salix* forests, San Pedro River, Arizona. Biol Conserv 125(2):153–167.

Logan, JA; Powell, JA. (2001) Ghost forests, global warming, and the mountain pine beetle (*Coleoptera: Scolytidae*). Am Entomol 47(3):160–172.

Lonsdale, WM. (1993) Rates of spread of an invading species – *Mimosa pigra* in northern Australia. J Ecol 81(3):513–521.

Lonsdale, WM. (1999) Global patterns of plant invasions and the concept of invasibility. Ecology 80(5):1522–1536.

Mack, RN; Simberloff, D; Lonsdale, WM; et al. (2000) Biotic invasions: causes, epidemiology, global consequences, and control. Ecol Appl 10(3):689–710.

MacLean, D. (2007) Status of state ANS management plans. U.S. Fish and Wildlife Service, Aquatic Nuisance Species Task Force. Available online at http://www.anstaskforce.gov/Documents/StatePlansAug2007.pdf.

Mandrak, N.E. (1989) Potential invasion of the Great Lakes by fish species associated with climatic warming. J Great Lakes Res 15(2):306–316.

Masifwa, WF; Twongo, T; Denny, P. (2001) The impact of water hyacinth, *Eichhornia crassipes* (Mart) Solms on the abundance and diversity of aquatic macroinvertebrates along the shores of northern Lake Victoria, Uganda. Hydrobiologia 452(1–3):79–88.

Massachusetts Aquatic Invasive Species Working Group. (2002) Massachusetts aquatic invasive species management plan. Massachusetts Office of Coastal Zone Management. Available online at: http://www.anstaskforce.gov/Mass_AIS_Plan.pdf.

McCarty, J. (2001) Ecological consequences of recent climate changes. Conserv Biol 15(2):320–331.

McFarland, DJ; Barko, JW. (1999) High-temperature effects on growth and propagule formation in hydrilla biotypes. J Aquat Plant Manage 37:17–35.

McLaughlin, JF; Hellmann, JJ; Boggs, CL; et al. (2002) Climate change hastens population extinctions. Proc Natl Acad Sci USA 99(9):6070–6074.

MEA (Millennium Ecosystem Assessment). (2005) Ecosystems and human well-being. Volume 1: current state and trends. Hassan, R; Scholes, R; Ash, N; eds. Washington, DC: Island Press.

Meacham, P. (2001) Washington aquatic nuisance species management plan. Washington Department of Fish and Wildlife. Available online at http://wdfw.wa.gov/fish/ans/2001ansplan.pdf.

Melbourne, BA; Cornell, HV; Davies, KF; et al. (2007) Invasion in a heterogeneous world: resistance, coexistence or hostile takeover? Ecol Lett 10(1):77–94.

Michener WK; Blood, ER; Bildstein, KL; et al. (1997) Climate change, hurricanes and tropical storms, and rising sea level in coastal wetlands. Ecol Appl 7(3):770–801.

Miller, RR; Williams, JD; Williams, JE. (1989) Extinctions of North-American fishes during the past century. Fisheries 14(6):22–38.

Mooney, HA; Hobbs, RJ; eds. (2000) Invasive species in a changing world. Washington, DC: Island Press.

Mueller, JM; Hellmann JJ. (2008) An assessment of invasion risk from assisted migration. Cons Biol 22: in press.

Nelson, GC. (2005) Drivers of ecosystem change: summary chapter. In: Hassan, R; Scholes, R; Ash, N; eds. Ecosystems and human well-being. Volume 1: current state and trends. Washington, DC: Island Press; pp. 73–76.

NISC (National Invasive Species Council). (2001) National invasive species management plan: meeting the invasive species challenge. United States Department of Agriculture. Available online at http://www.invasivespeciesinfo.gov/council/nmptoc.shtml.

NISIC (National Invasive Species Information Center). (2006) Resource library. United States Department of Agriculture. Available online at http://www.invasivespeciesinfo.gov/resources/orgcouncilstate.shtml [accessed September, 21 2006].

Novacek MJ; Cleland EE. (2001) The current biodiversity extinction event: scenarios for mitigation and recovery. Proc. Natl Acad Sci 98(10):5466–5470.

Ojala, A; Kankaala, P; Tulonen, T. (2002) Growth response of *Equisetum fluviatile* to elevated CO_2 and temperature. Environ Exp Bot 47(2):157–171.

OTA (Office of Technology Assessment). (1993) Harmful non-indigenous species in the United States. U.S. Congress Office of Technology Assessment; OTA-F-565. Available online at http://govinfo.library.unt.edu/ota/Ota_1/DATA/1993/9325.PDF.

Parmesan, C. (2006) Ecological and evolutionary responses to climate change. Annu Rev Ecol Evol Syst 37:637–69.

Pearson, RG; Dawson, TP. (2003) Predicting the impacts of climate change on the distribution of species: are bioclimate envelope models useful. Global Ecol Biogeogr 12(5):361–371.

Pelejero, C; Calvo, E; McCulloch, MT; et al. (2005) Preindustrial to modern interdecadal variability in coral reef pH. Science 309(5744):2204–2207.

Peterson, AT. (2003) Predicting the geography of species' invasions via ecological niche modeling. Q Rev Biol 78(4):419–433.

Peterson, AT; Vieglais, DA. (2001) Predicting species invasions using ecological niche modeling: new approaches from bioinformatics attack a pressing problem. Bioscience 51(5):363–371.

Peterson, AT; Ortega-Huerta, MA; Bartley, J; et al. (2002) Future projections for Mexican faunas under global climate change scenarios. Nature 416(6881):626–629.

Phelps, HL. (1994) The asiatic clam (*Corbicula fluminea*) invasion and system-level ecological change in the Potomac River estuary near Washington, DC. Estuaries 17(3):614–621.

Pimentel, D. (2003) Economic and ecological costs associated with aquatic invasive species: proceedings of the aquatic invaders of the Delaware estuary symposium; May 20, 2003; Penn State Great Valley Campus, Malvern,

PA; pp.3–5. Available online at
http://sgnis.org/publicat/proceed/aide/Aquatic%20Invaders%20of%20the%20Delaware%20Esutary.pdf.

Pimentel, D; Lach, L; Zuniga, R; et al. (2000) Environmental and economic costs of nonindigenous species in the United States. BioScience 50(1):53–65.

Pimentel, D; Zuniga, R; Morrison, D. (2005) Update on environmental and economic costs associated with alien-invasive species in the United States. Ecol Econ 52:273–288.

Porter, WP; Sabo, JL; Tracy, CR; et al. (2002) Physiology on a landscape scale: plant-animal interactions. Integ Comp Biol 42(5):431–453.

Pyke, CR; Thomas, R; Porter, RD; et al. (2008) Current practices and future opportunities for climate change and invasive species policy. Conserv Bio: in press.

Rahel, FJ; Olden, JD. (2008) Assessing the effects of climate change on aquatic invasive species. Conserv Bio: in press.

Ramey, V. (2001) Non-native invasive aquatic plants in the United States: *Eichhornia crassipes*. Center for Aquatic and Invasive Plants, University of Florida and Seagrant. Available online at http://aquat1.ifas.ufl.edu/seagrant/eiccra2.html [accessed May 2007].

Rejmánek, M; Richardson, DM. (1996) What attributes make some plant species more invasive? Ecology 77(6):1655–1661.

Ricciardi A. (2001) Facilitative interactions among aquatic invaders: is an 'invasional meltdown' occurring in the Great Lakes? Can J Fish Aquat Sci 58:2513–2525.

Ricciardi, A. (2006) Patterns of invasion in the Laurentian Great Lakes in relation to changes in vector activity. Divers Distrib 12(4):425–433.

Richardson, DM; Pyšek, P. (2006) Plant invasions – merging the concepts of species invasiveness and community invasibility. Prog Phys Geog 30(3):409–431.

Root, TL; Price, JT; Hall, KR; et al. (2003) Fingerprints of global warming on wild animals and plants. Nature 421(6918):57–60.

Roura-Pascual, N; Suarez, AV; Gomez, C; et al. (2004) Geographical potential of Argentine ants (*Linepithema humile* Mayr) in the face of global climate change. Proc R Soc Lond B 271(1557):2527–2534.

Ruiz, GM; Carlton, JT. (2003) Invasion vectors: conceptual framework for management. In Ruiz, GM; Carlton, JT; eds. Invasive species: vectors and management strategies. Washington, DC: Island Press; pp. 459–504.

Scavia, D; Field, JC; Boesch, DF; et al. (2002). Climate change impacts on U.S. coastal and marine ecosystems. Estuaries 25(2):149–164.

Schnitzler, A; Muller, S. (1998) Ecology and biogeography of highly invasive plants in Europe: giant knotweeds from Japan (*Fallopia japonica* and *F. sachalinensis*). Rev Ecol Terre Vie 53(1):3–38.

Seager, R; Ting, M; Held, I; et al. (2007) Model projections of an imminent transition to a more arid climate in southwestern North America. Science 316(5828):1181–1184.

Shafroth, PB; Stromberg, JC; Patten, DT. (2002) Riparian vegetation response to altered disturbance and stress regimes. Ecol Appl 12(1):107–123.

Shluker, AD. (2003) State of Hawaii aquatic invasive species management plan. Department of Land and Natural Resources – Division of Aquatic Resources. Available online at http://hawaii.gov/dlnr/dar//pubs/ais_mgmt_plan_final.pdf.

Siguan, MAR. (2003) Pathways of biological invasions of marine plants. In: Ruiz, GM; Carlton, JT; eds. Invasive species: vectors and management strategies. Washington, DC: Island Press; pp. 183–226.

Simberloff, D. (2003) Eradication-preventing invasions at the outset. Weed Sci 51(2):247–253.

Simberloff, D; Von Holle, B. (1999) Positive interactions of nonindigenous species: invasional meltdown? Biol Invas 1:21–32.

Simon, KS; Townsend, CR. (2003) Impacts of freshwater invaders at different levels of ecological organisation, with emphasis on salmonids and ecosystem consequences. Freshwat Biol 48:982–994.

Stachowicz, JJ; Terwin, JR; Whitlatch, RB; et al. (2002) Linking climate change and biological invasions: ocean warming facilitates nonindigenous species invasions. Proc Natl Acad Sci USA 99(24):15497–15500.

Stewart, IT; Cayan, DR; Dettinger, MD. (2004) Changes in snowmelt runoff timing in western North America under a 'business as usual' climate change scenario. Clim Change 62:217–232.

Stireman, JO; Dyer, LA; Janzen, DH; et al. (2005) Climatic unpredictability and parasitism of caterpillars: implications of global warming. Proc Natl Acad Sci USA 102(48):17384–17387.

Stockwell DRB, Noble IR. (1992) Induction of sets of rules from animal distribution data: a robust and informative method of data analysis. Math Comp Simul 33(5–6):385–390.

Stockwell DRB, Peters D. (1999) The GARP modeling system: problems and solutions to automated spatial prediction. Int J Geo Info Sci 13(2):143–158.

Strayer, DL; Malcom, HM. (2006) Long-term demography of a zebra mussel (*Dreissena polymorpha*) population. Freshwat Biol 51(1):117–130.

Tinner, W; Lotter, AF. (2001) Central European vegetation response to abrupt climate change at 8.2 ka. Geology 29(6):551–554.

Toft, JD; Simenstad, CA; Cordell, JR; et al. (2003) The effects of introduced water hyacinth on habitat structure, invertebrate assemblages, and fish diets. Estuaries 26(3):746–758.

Tol, RSJ. (2002) Estimates of the damage costs of climate change. Part one: benchmark estimates. Environ Resour Econ 21:47–73.

Underwood, EC; Kilinger, R; Moore, P. (2004) Predicting patterns of non-native plant invasions in Yosemite National Park, California, USA. Divers Distrib 10(5–6):447–459.

USFWS (US Fish and Wildlife Service). (1996) Fact sheet: purple loosestrife. Available online at http://www.ceris.purdue.edu/napis/pests/pls/factspls.txt.

USFWS (US Fish and Wildlife Service). (2005) Planning is everything: managing natural resource pathways. Available online at http://www.haccp-nrm.org/Documents/PathwayMgtFactsheet2005.pdf [accessed January 8, 2007].

van Asch, M; Visser, ME. (2007) Phenology of forest caterpillars and their host trees: the importance of synchrony. Annu Rev Entomol 52:37–55.

Vitousek, PM. (1994) Beyond global warming: ecology and global change. Ecology 75(7):1861–1876.

Vitousek, PM; D'Antonio, CM; Loope, LL; et al. (1996) Biological invasions as global environmental change. Am Sci 84(5):468–478.

Vitousek, PM; D'Antonio, CM; Loope, LL. (1997a) Introduced species: a significant component of human-caused global change. N Z J Ecol 21(1):1–16.

Vitousek, PM; Mooney, HA; Lubchenco, J; et al. (1997b) Human domination of earth's ecosystems. Science 277(5325):494–499.

Walther, GR; Post, E; Convey, P; et al. (2002) Ecological responses to recent climate change. Nature 416(6879):389–395.

Walton, WC; MacKinnon, C; Rodriguez, LF; et al. (2002) Effect of an invasive crab upon a marine fishery: green crab, *Carcinus maenas*, predation upon a venerid clam, *Katelysia scalarina*, in Tasmania (Australia). J Exp Mar Biol Ecol 272(2):171–189.

Weltzin, JF; Belote, TR; Sanders, NJ. (2003) Biological invaders in a greenhouse world: will elevated CO_2 fuel plant invasions? Front Ecol Environ 1(3):146–153.

Wilcox, KL; Petrie, SA; Maynard, LA; et al. (2003) Historical distribution and abundance of *Phragmites australis* at Long Point, Lake Erie, Ontario. J Great Lakes Res 29(4):664–680.

Winder, M; Schindler, DE. (2004a) Climatic effects on the phenology of lake processes. Global Change Biol 10(11):1844–1856.

Winder, M; Schindler, DE. (2004b) Climate change uncouples trophic interactions in an aquatic ecosystem. Ecology 85(8):2100–2106.

Wittenberg, R; Cock, MJW. (2001) Invasive alien species: how to address one of the greatest threats to biodiversity: a toolkit of best prevention and management practices. Wallingford, Oxon, UK: CAB International.

Zedler, JB; Kercher, S. (2004) Causes and consequences of invasive plants in wetlands: opportunities, opportunists, and outcomes. Crit Rev Plant Sci 23(5):431–452.

Ziska, LH. (2003a) Evaluation of yield loss in field sorghum from a C_3 and C_4 weed with increasing CO_2. Weed Sci 51(6):914–918.

Ziska, LH. (2003b) Evaluation of the growth response of six invasive species to past, present and future atmospheric carbon dioxide. J Exp Biol 54(381):395–404.

Ziska, LH; Teasdale, JR; Bunce, JA. (1999) Future atmospheric carbon dioxide may increase tolerance to glyphosate. Weed Science 47(5) 608–615.

Ziska, LH; George, K; Frenz, DA. (2007) Establishment and persistence of common ragweed (*Ambrosia artemisiifolia* L.) in disturbed soil as a function of an urban-rural macro-environment. Global Change Biol 13(1):266–274.

APPENDIX A

AQUATIC INVASIVE SPECIES PROGRAMS AND ACTIVITIES 50-STATE SUMMARY

CONTENTS

CONTENTS (continued)

CONTENTS (continued)

A.1. METHODS

We inventoried aquatic invasive species (AIS)-related management actions in all 50 states to determine what information may be needed to allow AIS managers to consider and incorporate predicted global change impacts into their programs. For each state, we documented the status of AIS management plans, state programs and activities, climate change concerns, climate change actions, and research activities and needs. We reviewed publicly available documents, state agency publications, and online materials. For further clarification, when appropriate, staff from the Environmental Law Institute (ELI)-discussed AIS programs, research needs, and management strategies with AIS managers, scientists, and decision makers. Each state summary was sent to both state agency and U.S. Environmental Protection Agency regional staff for review and comment in November and December of 2006. Comments were vetted, and summaries finalized, in January 2007. Note—State plans generally refer to AIS as aquatic nuisance species or ANS.

A.2. SUMMARY OF AQUATIC INVASIVE SPECIES MANAGEMENT IN ALABAMA

A.2.1. AQUATIC INVASIVE SPECIES MANAGEMENT PLAN

Plan under development.

A.2.2. AQUATIC INVASIVE SPECIES PROGRAMS AND ACTIVITIES

- **Aquatic Plant Management Control Program, Alabama Department of Conservation and Natural Resources (AL DCNR), Division of Wildlife and Freshwater Fisheries (DWFF), and U.S. Army Corps of Engineers, Mobile District.** The Program conducts surveys to determine presence of aquatic nuisance plants and control for aquatic nuisance plants using herbicides.

- **Private Waters, AL DCNR, and DWFF.** The program provides technical guidance to private pond owners for aquatic invasive species (AIS) removal.

- **Mobile Bay National Estuary Program, Alabama-Mississippi Rapid Assessment Team.** This program conducts a 3–5-day surveys of all aquatic, invasive species present in the coastal waters of Alabama and Mississippi to establish a baseline. The program was launched in 2003 with 50 scientists surveying Mobile Bay and targeting the Mississippi Sound and adjacent waters. The 2004 survey was conducted by more than 100 scientists from 26 organizations and constituted the largest rapid assessment of living resources ever held in the Gulf of Mexico.

A.2.3. CLIMATE CHANGE CONCERNS

- Alabama has experienced a lack of a cold winters in recent years, which may or may not be attributed to climate change. These warmer winters have allowed invasive plants and fish (e.g., Nile tilapia) to overwinter and to move further north.

A.2.4. CLIMATE CHANGE ACTIONS

(None reported.)

A.2.5. RESEARCH ACTIVITIES AND INFORMATION USED

- In determining where to undertake control work, the AL DCNR and DWFF look for areas with significant impacts to fisheries, as well as areas with detrimental impacts to boating access and angler usage. The identification of survey areas is based on prior existence of plant problems. Areas with a history of plant problems are included, while areas with no past history of plant problems are excluded.

A.2.6.   RESEARCH NEEDS

- AIS management in Alabama needs more effective herbicides, with better long-term control.

- There is also a need for more information and an enhanced strategy for emergent control.

- In addition, AIS management would benefit from surveys conducted by experts on non-native species, as well as fund to secure these services.

A.3. SUMMARY OF AQUATIC INVASIVE SPECIES MANAGEMENT IN ALASKA

A.3.1. STATUS OF AQUATIC INVASIVE SPECIES MANAGEMENT PLAN

Alaska's Aquatic Nuisance Species (ANS) Management Plan was published in 2002 (see Appendix B, State Aquatic Invasive Species Management Plan Summaries for a general description of the Plan).

A.3.2. AQUATIC INVASIVE SPECIES PROGRAMS AND ACTIVITIES

- **Kenai Peninsula Cooperative Weed Management Area (CWMA), Homer Soil and Water Conservation District (SWCD), Alaska SWCD, Kenai SWCD.** The SWCDs have established an advisory board and listed priorities for the Weed Management Area.

- **Noxious and Invasive Plant Program, Upper Susitna SWCD.** This program targets the local airport to prevent the transport (airplanes, luggage, and shoes) and spread of Orange hawkweed. Other activities include herbicide application and volunteer weed pulling in cooperation with the University of Alaska-Fairbanks' Cooperative Extension Service.

- **Weed Ranking Program, Alaska Natural Heritage Program, University of Alaska, Anchorage, Environmental and Natural Resources Institute.** The Alaska Natural Heritage Program, in cooperation with other federal and state agencies, developed the Weed Ranking Project, which lists and ranks non-native plant species.

- **Alaska Exotic Plant Information Clearing House (AK EPIC) Mapping Project, Alaska Natural Heritage Program, University of Alaska, Anchorage, Environmental and Natural Resources Institute.** The Alaska Natural Heritage Program also partners with the U.S. Department of Agriculture's Forest Service/State and Private Forestry Service, the National Park Service, and the U.S. Geological Survey, Alaska Science Center on the AK EPIC. The AK EPIC draws much of its information from surveys, encompasses data from CWMAs, and employs a rapid response program.

- **Alaska Committee for Noxious and Invasive Plants Management, University of Alaska, Fairbanks, Cooperative Extension Service.** This committee was established in 2003 to encourage and work towards a coordinated, statewide effort to prevent, manage, and increase the awareness of invasive and noxious species.

- **Alaska Invasive Species Working Group.** This group was formed in 2006 to work towards an all-taxa, statewide invasive species cooperative effort. Members include state, federal, non-governmental organizations, and Alaska Native organizations. The group is currently working on an Alaska invasive species needs assessment.

- **Northern Pike Education Program, Alaska Department of Fish and Game (AK DFG), Sport Fish Division.**

- **Kachemak Bay Research and Reserve Green Crab Community Monitoring Program, Prince William Sound Science Center, AK DFG, National Oceanic and Atmospheric Administration (NOAA), and local communities.** This program provides a protocol for schoolchildren to learn the biology of green crabs in order to do monitoring work.

A.3.3.   CLIMATE CHANGE CONCERNS

- Alaska's aquatic invasive species (AIS) plan predicts an increase in invasive species as warmer temperatures allow for overwintering. Species of concern include: the mitten crab, yellow perch, and walleyed pike.

- The state is conducting a risk assessment study for the Chinese mitten crab, because climate change will most likely result in the migration of this species to Alaska.

- Water temperatures have warmed to the point where shellfish could survive through the winter season, resulting in a shellfish outbreak.

- State officials are also concerned with species moving from one part of the state to another.

A.3.4.   CLIMATE CHANGE ACTIONS

- Alaska's ANS Management Plan focuses on prevention and identification of the most prominent threats. The management plan recognizes that the southern areas with "warmer climate, more developed lands, more disturbed habitat, and better road access" are areas of particular concern. Alaska's ANS plan identifies ports with high traffic as posing greater risk.

- The Weed Ranking Project provides a way to prioritize work. The project ranks both non-native species present in the state and those species that are likely to invade Alaska due to climate change. A "climate match" program loosely associates species with one of Alaska's ecosystems (maritime, boreal, or arctic) to address these concerns.

A.3.5.   RESEARCH ACTIVITIES AND INFORMATION USED

- Regional Alaskan groups are monitoring for green crab and, where found, setting traps as a control method.

- Proposals for mapping and inventorying of reed canary grass have been developed.

- Research on the effects of rats on the ecosystem through local projects and case studies, including examining the effects of rats on intertidal invertebrates and soil composition and testing rodenticides is needed.

- State officials are inventorying all exotic plant species. This collection includes about 130 species, of which approximately 20 are expected to be a problem. Of these 20, only a few are found in riparian areas.

- Statewide northern pike management plan, which was expected to be completed by the end of 2006 by AK DFG, and the Upper Susitna/Copper River Pike Surveys to determine how widespread pike are in the area.

- Ballast water-related research will be funded in FY07/08 by NOAA Sea Grant and administered by AK DFG.

- Risk assessment for aquatic sea lice will be funded in FY07/08 by NOAA Sea Grant and administered by AK DFG.

- Ongoing shore zone mapping research will characterize the physical and biological attributes of each section of the shoreline.

- Activities also include ranking the invasiveness of non-native animals and fish.

A.3.6. RESEARCH NEEDS

- New and improved pike control techniques are needed. Ideally, a piscicide would work best. Options to control pike are currently limited to netting and four approved chemicals (Rotenone, antimycin, TFM, and Bayluscide).

- Alaskan AIS management would benefit from the development of aquaculture systems that will prevent salmon escape.

- Additional knowledge about the speed at which green crabs are entering the state is needed. In general, this species moves slowly, but officials must learn more about its migration in order to determine the scope of any potential problems.

- In order to address the migration of green crabs, Alaska needs to develop pheromones and trapping methods. Green crab trapping methods also require more in-depth research that addresses questions such as: Is it possible to develop techniques to trap them out completely? What are the best techniques for managing them at a low level, with compounds that will attract them quickly into traps? Also, what is the ideal type of trap?

- Research on the locations of green crabs could also provide the state with a better understanding of the species' different ecological needs.

- The mechanisms for how reed canary grass affects water quality need to be better understood.

- To prevent invasion of colonial tunicates, pathways need to be better understood.

A.4. SUMMARY OF AQUATIC INVASIVE SPECIES MANAGEMENT IN ARIZONA

A.4.1.   AQUATIC INVASIVE SPECIES MANAGEMENT PLAN

Arizona's aquatic invasive species (AIS) management plan is under development (see Appendix B, State Aquatic Invasive Species Management Plan Summaries for a general description of the Plan).

A.4.2.   AQUATIC INVASIVE SPECIES PROGRAMS AND ACTIVITIES

- **Invasive Species Council, Arizona Game and Fish Department (AZGFD), Arizona Department of Agriculture (DA).** The council conducts a "Stop Aquatic Hitchhikers Program" and works with 100th Meridian, an initiative to stop the spread of zebra mussels, to inform watercraft operators/owners and marina operators to take proper precautions. The Council also conducts aquatic nuisance species (ANS) monitoring.

- **Giant Salvinia Task Force (GSTF), U.S. Bureau of Land Management, U.S. Bureau of Reclamation, U.S. Fish and Wildlife Service (USFWS), AZGFD, AZDA, California Department of Fish and Game, California Department of Food and Agriculture, Palo Verde Irrigation District, and 11 others.** Each of the 20 weed management area groups is responsible for implementing control efforts at a particular region. This particular Task Force has used intensive inventory, mechanical control, and herbicide application since 2001. Biocontrol (Salvinia weevils) was implemented in 2004 and has been followed by supplemental releases. An early detection and rapid response program is in place for invasive aquatic plants (e.g., a rapid response was undertaken recently for water hyacinth). This AIS Program works closely with the International Boundary and Water Commission.

- **Hydrilla Eradication, AZDA.** The AZDA and Arizona landowners continue treatment of two isolated populations of hydrilla in the Phoenix and Tucson areas as part of the regular enforcement of the state's noxious weed laws.

A.4.3.   CLIMATE CHANGE CONCERNS REPORTED BY STATE PERSONNEL

- It is generally accepted that climate has a relationship to the distribution of species, natural or introduced, and that the state needs to anticipate ecosystem changes as a result of changes in water temperature and environmental conditions.

- The Arizona State Wildlife Action Plan recognizes both climate change and invasive species as identified threats. As plant populations increase heavily during the summer, warmer temperatures due to climate change may generate more plant growth.

A.4.4. CLIMATE CHANGE ACTIONS

- The "Stop Aquatic Hitchhikers Program" and the 100th Meridian both inform watercraft operators/owners and marina operators to take proper precautions against AIS.

A.4.5. RESEARCH ACTIVITIES AND INFORMATION USED

- Animal and Plant Health Inspection Services has conducted a programmatic environmental assessment for the weevil.

- U.S. Department of Agriculture has used research carried out by the University of Arizona on new attempts at biocontrol and methods currently employed in other countries.

- USFWS, U.S. Bureau of Reclamation, and AZGFD have sponsored preliminary investigations into genetic biocontrol. AZGFD and AZDA have also conducted some monitoring.

- The GSTF is monitoring the spread of giant salvinia and attempting to document efficacy.

A.4.6. RESEARCH NEEDS

- Effective control methods for crayfish need to be developed. The University of Arizona is undertaking research into crayfish life histories in an effort to identify vulnerabilities for population control.

- Advantages and disadvantages of biological, mechanical, and chemical control options need to be determined for hydrilla, salvinia, and other aquatic nuisance plants.

- Information on how to coordinate activities of multiple state agencies with overlapping jurisdiction needs to be gathered.

- Research the effectiveness of weevils for biocontrol is needed, though this is hampered by a lack of funding.

A.5. SUMMARY OF AQUATIC INVASIVE SPECIES MANAGEMENT IN ARKANSAS

A.5.1. AQUATIC INVASIVE SPECIES MANAGEMENT PLAN

Arkansas's aquatic invasive species (AIS) management plan is under development.

A.5.2. AQUATIC INVASIVE SPECIES PROGRAMS AND ACTIVITIES

- **Noxious Weed Programs (Purple Loosestrife, Giant Salvinia, and Water Hyacinth), Arkansas State Plant Board.** The board implements regulations pertaining to invasive species.

- **Hydrilla Control, State of Arkansas in cooperation with the U.S. Army Corps of Engineers.** At Lake Ouachita, officials are trying to reduce the infestation by providing grass carp as a biocontrol.

- **Arkansas River Study, Arkansas Game and Fish Commission.** This study involves ongoing large river sampling of many species, including Asian carp.

A.5.3. CLIMATE CHANGE CONCERNS

- Officials believe that new invasive species will survive the winter and persist in the state. Species already established may be allowed to spread into northern areas.

- Invasive species may enter the state as a result of increased interstate commerce and boating.

A.5.4. CLIMATE CHANGE ACTIONS

- Arkansas recently enacted regulations targeting water hyacinth due to overwintering concerns.

- Arkansas is formulating a state AIS management plan that will include measures that address warming air and water temperatures.

- Regulate the aquaculture industry.

A.5.5. RESEARCH ACTIVITIES AND INFORMATION USED

- Purple loosestrife and giant salvinia surveys are needed.

- Zebra mussels, hydrilla, and Asian carp need to be monitored and their occurrence and magnitude of infestation need to be documented.

A.5.6. RESEARCH NEEDS

- Information on the Asian carp, including its abundance, impacts, and pathways, is needed.

- Information on the zebra mussels, including its impacts and pathways, is needed.

- Information about species that may potentially enter the state as a result of interstate commerce is needed.

A.6. SUMMARY OF AQUATIC INVASIVE SPECIES MANAGEMENT IN CALIFORNIA

A.6.1. AQUATIC INVASIVE SPECIES MANAGEMENT PLAN

California Aquatic Invasive Species (AIS) Management Plan was approved in November 2007.

A.6.2. AQUATIC INVASIVE SPECIES PROGRAMS AND ACTIVITIES

- **Aquatic Pest Control Program, California Department of Boating and Waterways (CADBW).** This Program focuses on control of water hyacinth, *Caulerpa taxifolia*, and *Egeria densa*. The CADBW also uses annual hyperspectral aerial survey to monitor changes in infestations over time. The CADBW uses short- and long-term methods of water hyacinth control, involving chemical, mechanical, and biocontrol measures. The department also works with the California Department of Fish and Game on *Caulerpa* eradication efforts in southern California under the direction of the Southern California Caulerpa Action Team. Officials are also trying to educate aquarium owners on *Caulerpa*. The *Egeria densa* Control Program for the Delta focuses mainly on herbicide control.

- **Hydrilla Eradication Program, California Department of Food and Agriculture, Integrated Pest Control Branch.** The program conducts annual surveys and eradication efforts for hydrilla. Eradication consists of physical, biological, and chemical methods.

A.6.3. CLIMATE CHANGE CONCERNS

- Aquarium owners may serve as vectors for *Caulerpa* spread.

- Increased interstate transport as a result of climate change, such as opening of previously frozen waterways, may enable spread of AIS.

A.6.4. CLIMATE CHANGE ACTIONS

- Annual hyperspectral aerial survey should be used to monitor changes in infestations over time.

- Educate aquarium owners about *Caulerpa*.

- Implement prevention methods including quarantine regulations, inspection programs to ensure compliance with quarantine regulations, and border inspection stations to screen incoming traffic.

- Work to detect invasive species using insect traps, manual inspections for exotic weed species, and/or surveys to determine size and boundaries of population.

A.6.5. RESEARCH ACTIVITIES AND INFORMATION USED

- Annual hyperspectral aerial surveys are being used to monitor changes in infestations over time.

- Annual surveys for hydrilla are taking place.

- Invasions are being monitored with insect traps and manual inspections.

A.6.6. RESEARCH NEEDS

- Additional research on the biology/DNA of *Caulerpa* and how it would adapt in Southern California, as well as research on eradication methods is needed. Officials would also like to undertake greater surveillance.

- Additional outreach and public education regarding *Caulerpa* also is needed. Individuals (hobbyists) need to learn how to handle *Caulerpa* (it is important to teach people how to look out for it in the natural environment).

A.7. SUMMARY OF AQUATIC INVASIVE SPECIES MANAGEMENT IN COLORADO

A.7.1. AQUATIC INVASIVE SPECIES MANAGEMENT PLAN

No plan available.

A.7.2. AQUATIC INVASIVE SPECIES PROGRAMS AND ACTIVITIES

- **Aquatic Plants Management Program, Colorado Department of Agriculture (CODA).** The program operates several projects throughout the state. Work includes both manual removal and chemical treatments. Presently, the focus is on the Rio Grande Watershed, the upper part of the Colorado River, the North Platte River, the San Miguel River, and the Republican River watershed. (The main coordinator for the San Miguel Project is The Nature Conservancy.) There are also control efforts under way for Siberian Elm, including mechanical removal, herbicide application, and cut stump treatment. There are plans to implement biocontrol for tamarisk as well.

- **Biocontrol of Tamarisk, CODA.** The Department's Insectary in Palisade, Colorado is the clearinghouse for the project. Officials are working in collaboration with U.S. Department of Agriculture and Colorado State University to release beetles in Colorado, Wyoming, South Dakota, Montana, Oregon, Kansas, and Idaho to control tamarisk. About 60,000 tamarisk leaf beetles have been released in seven states with additional releases planned. In August 2005, beetles were released at three Colorado sites: Adams, Mesa, and Yuma Counties. In 2006, beetles were released at Dinosaur National Monument in Moffat County and several additional sites in Colorado and the West.

- **Aquatic Animal Management Program, Department of Natural Resources, Division of Wildlife (CODW).** Major activities of CODW on aquatic invasive species (AIS) include: (1) angler education; (2) hatchery maintenance; (3) activities to detect location of New Zealand mud snails (NZMS); and (4) participation in the Western Regional Panel of the Aquatic Nuisance Species (ANS) Task Force. Colorado State Parks is cooperating with CODW by providing them with GIS/GPS training, ANS mapping access/support, and collaborating on various education projects, control methods, and statewide planning efforts.

- **Eurasian Watermilfoil (EWM) activities, Colorado State Parks.** The Stewardship Section of Colorado State Parks is the central coordinator and GIS clearinghouse for EWM efforts in Colorado. The program is actively working towards several short- and long-term objectives that include coordination, mapping, data collection, grant writing, planning, early detection and rapid response, partnering with local universities on research, education campaigns, convening stakeholders, studying economic impacts, and implementation of boat washing stations.

A.7.3. CLIMATE CHANGE CONCERNS

- Species that currently cannot overwinter in Colorado, such as giant salivinia or water hyacinth, may persist if climate changes occur and water temperatures increase. This depends both on whether the water is hot or spring fed and on the location of the species within the state.

A.7.4. CLIMATE CHANGE ACTIONS

- Angler education program focuses on prevention through outreach, including posting angler alert signs at trout fishing locations and live fishing tackle stores.

- Hatchery maintenance program ensures that fish production units remain free of invasive species.

A.7.5. RESEARCH ACTIVITIES AND INFORMATION USED

- Colorado is involved with the tamarisk biocontrol program. One of the first field sites used to test biocontrol beetles in North America was located near Pueblo, Colorado. The CODA Insectary has been involved in the project for several years and has received a permit to store up to 1 million beetles for use in biocontrol in 2005.

- Weed researchers in the state are studying AIS and the use of biocontrol. These researchers are collaborating with federal agencies such as U.S. Geological Survey.

- Records on the location of newly discovered species are being maintained.

A.7.6. RESEARCH NEEDS

(Numerous research needs, but none specifically provided.)

A.8. SUMMARY OF AQUATIC INVASIVE SPECIES MANAGEMENT IN CONNECTICUT

A.8.1. AQUATIC INVASIVE SPECIES MANAGEMENT PLAN

The Connecticut Aquatic Nuisance Species (ANS) Management Plan was approved in 2006 (see Appendix B, State Aquatic Invasive Species Management Plan Summaries for a general description of the Plan).

A.8.2. AQUATIC INVASIVE SPECIES PROGRAMS AND ACTIVITIES

- **Non-Native Invasive Plant Species Program, Connecticut Department of Environmental Protection (CT DEP).** The program conducts the following activities: (1) *Rapid response and eradication* of newly-introduced aquatic plants, including water chestnut (the CT DEP Fisheries Division and Wildlife Division and the Office of Long Island Sound are working with U.S. Fish and Wildlife Service (USFWS) to carry out eradication projects and surveys); (2) *Restoration* of coastal habitats, including Phragmites control in saltwater tidal marshes; and (3) *Implementation of a Rapid Response Plan* for *Hydrilla verticillata* that was prepared by Wildlife Division staff.

- **Invasive Plant Council (IPC).** The IPC was established in 2003 under Connecticut state law. Membership includes representatives from state agencies, universities, Invasive Plant Atlas of New England (IPANE), non-profit conservation groups, and the Connecticut Nurseryman's Association. To date, a total of 81 non-native invasive plant species have been listed with prohibitions on importation, moving, sale, purchase, transplantation, cultivation, or distribution. The IPC is currently working on obtaining funding to create a non-native invasive plant program that will focus on early detection, rapid response, education, and prevention.

- **Wetlands Habitat and Mosquito Management (WHAMM) Program, CT DEP, Wildlife Division.** The control of *Phragmites australis* through herbicide application has been a major component of recent wetland restoration efforts. The WHAMM Program also plans to research new alternative herbicides for Phragmites control.

- **Water Chestnut Harvesting Program, CT DEP, Fisheries Division and Wildlife Division, Office of Long Island Sound Programs, USFWS Connecticut River Coordinator's Office (Connecticut River Fisheries Program).** The program conducts water chestnut management (surveys, removal, education) and monitoring for undiscovered water chestnut populations.

- **Lakes Management Program, CT DEP, Bureau of Water Management, Division of Planning and Standards.** The program conducts the following activities: dredging of Silver Lake in Meriden/Berlin to hinder growth of Eurasian watermilfoil (EWM); funding to control variable watermilfoil in Bashan Lake with 2-4D; spot-treatment of EWM with limited amounts of herbicides; inventory and vegetation surveys of aquatic invasive plants, including listing of management options; and partnerships with communities to perform winter draw down, dredging, weed harvesting, and herbicide use.

- **IPANE, New England Wild Flower Society, University of Connecticut, Silvio O. Conte National Fish and Wildlife Refuge.** IPANE's mission is to create a comprehensive, continually updated web-accessible database of invasive and potentially invasive plants in New England. A network of professionals and trained volunteers will update the database, which will facilitate education and research that will lead to a greater understanding of invasive plant ecology and support informed conservation management. An important focus of the project is the early detection of, and rapid response to, new invasions.

- **Connecticut Sea Grant College Program, Sea Grant, University of Connecticut.** The program is working with Connecticut and New York agencies and organizations to develop an ANS Management Plan for the Long Island Sound, working with CT DEP to develop a state aquatic invasive species management plan, conducting outreach and education, participating on the Northeast ANS regional panel, and supporting research on red alga (*Grateloupia turuturu*), colonial tunicate (*Didemnum sp*), baitworms, the associated packing materials, and the economic impact of fouling organisms on marine aquaculture operations.

- **The Silvio O. Conte National Fish and Wildlife Refuge Invasive Plant Control Initiative.** The Refuge developed an Invasive Plant Control Initiative in response to the threat to natural diversity posed by invasive plant species. This initiative examines the problem of freshwater invasive plants from a regional perspective and identifies tasks that will enhance the capability within the region to address identified issues. Also, in cooperation with a number of partners, the Refuge used a grant from the National Fish and Wildlife Foundation to develop a strategic plan discussing the current invasive plant situation, outlining future actions for the Connecticut River Watershed and Long Island Sound, and recommending funding for high-priority invasive plant control projects in 1998. As part of the initiative, a partnership of federal, state, municipal, business, and non-profit groups formed to control water chestnut, a recent invader to the watershed. Components of the strategy include mechanical harvesting of the source population and organizing volunteers to monitor water bodies for satellite populations within the watershed and to hand-pull populations when found.

- **Research, Connecticut Agricultural Experiment Station (CTAES).** CTAES is researching control methods for nuisance aquatic plants, mapping their distribution and documenting the water conditions in which they are likely to occur. Studies are being conducted on control with herbicides and the effects of these products on nontarget plants. Water samples from treatment sites are being tested for herbicides to determine how concentrations change with time, where the herbicide may migrate, and what concentrations are necessary to achieve control with minimal impacts on desirable plants. Water from nearby wells is often tested to determine if aquatic herbicides can contaminate groundwater. Studies on the effectiveness of mechanical removal by different methods, including hydroraking and cutting, are also in progress. Biocontrol strategies, including studies on the distribution and preferences of the milfoil weevil (*Euhrychiopsis lecontei*) and a search for plant pathogens, are underway. A continuing statewide surveillance and mapping program of aquatic vegetation began in 2004. From 2004–2006, 126 lakes, including small private ponds, have been surveyed using global positioning system technology and GIS. Reference plants are being obtained from each water body and are being cataloged at CTAES herbaria and the University of Connecticut. Plant samples are also being frozen at -80°C for future molecular identification. Water chemistry and sediment data are being gathered from each lake to assess the preferences of nuisance plants and to determine the potential for other lakes to become infested.

A.8.3. CLIMATE CHANGE CONCERNS

- Residents release water hyacinth and water lettuce from their water gardens into state waters. With a warming trend, these species could overwinter and set seed, although there is no evidence of overwintering yet.

- A longer growing season could cause water chestnut to sprout earlier, persist longer into the fall, and produce more seeds. Water chestnut produce seeds more than once, flowering through the summer and fall before they start decomposing. A warmer climate would therefore make for a longer growing period. These plants might also grow faster with more light.

- There is a need for the development and update of lists of potential "new invaders." Early Detection and Rapid Response programs need to be developed and made operational for all taxonomic groups as the potential for new non-native invasive species may increase due to climatic changes.

A.8.4. CLIMATE CHANGE ACTIONS

- Coastal habitats are being restored (e.g., Phragmites control in saltwater tidal marshes). This restoration includes the re-establishment of tidal flows and the reintroduction of saltwater, both of which result in a gradual replacement of Phragmites by native vegetation.

A.8.5. RESEARCH ACTIVITIES AND INFORMATION USED

- Phragmites control methods include restoring tidal flows, mowing, herbicide application, and herbicide application with mowing, before selecting the herbicide glyphosate.

A.8.6. RESEARCH NEEDS

- For aquatic plants, a better systematic survey of the location of aquatic species in the state is needed, including assessment in small private ponds and trials on effective control methods for ANS.

- For water chestnut, a better understanding of the following is needed: germination of seeds based on temperature (whether a very cold winter would cause more seeds than usual to germinate at once the following spring); salinity limits; and biocontrols.

- Because correct identification of species is critically important to determining rapid response plans, there is a need for the development and use of genetic markers that will allow for positive identifications.

A.9. SUMMARY OF AQUATIC INVASIVE SPECIES MANAGEMENT IN DELAWARE

A.9.1. AQUATIC INVASIVE SPECIES MANAGEMENT PLAN

No plan available.

A.9.2. AQUATIC INVASIVE SPECIES PROGRAMS AND ACTIVITIES

- **Survey and inventory of aquatic vegetation in Delaware ponds, Department of Natural Resources and Environmental Control (DNREC), Division of Fish and Wildlife.** This program has two components: (1) control of aquatic nuisance species in public ponds, and (2) survey and mapping of aquatic vegetation in public ponds (invasive and rare species). Species surveyed and mapped range from open water species to the emergent shoreline vegetation. The department uses the maps to calculate the acreage figures, which can be used to document the species changes over time.

- **Delaware Landowner Incentive Program (DELIP), DNREC Division of Fish and Wildlife.** DELIP provides grant assistance to private landowners for habitat restoration, including invasive species control projects.

- **Phragmites Control Program, DNREC Division of Fish and Wildlife.** The program uses helicopter application of herbicides to control Phragmites in state wildlife areas and private lands (cost-share arrangement between landowners and the state).

- **Technical assistance to pond owners, DNREC Division of Fish and Wildlife.** The division provides assistance with invasive weed control, including recommendations on herbicides, manual control, or biocontrol, and dissemination of best management practices such as riparian buffer strips and nutrient control.

- **Delaware Invasive Species Tracking System, Delaware Natural Heritage Program, Delaware Invasive Species Council, U.S. Geological Survey, Leetown Science Center.** The system is a prototype for invasive species reporting and tracking. The goal is to develop an online tool for mapping and cataloging locations of invasive species in the state.

- **Wildlife Habitat Incentives Program, U.S. Department of Agriculture (USDA) Natural Resources Conservation Service, DNREC.** This is a cost-share program for private landowners who control Phragmites on their property. The DNREC provides a share of the cost (30%) and conducts the spraying. The USDA and the landowner also provide shares of the cost (58% and 12%), respectively.

- **Delaware River Invasive Plant Partnership, States of Delaware, New Jersey, New York, and Pennsylvania.**

A.9.3. CLIMATE CHANGE CONCERNS

(None reported.)

A.9.4. CLIMATE CHANGE ACTIONS

- DNREC surveys and maps species ranging from open water to emergent shoreline vegetation. The department uses these maps to calculate the acreage figures, which can be used to document species changes over time.

A.9.5. RESEARCH ACTIVITIES AND INFORMATION USED

- The Delaware Invasive Species Tracking System is being developed as an online tool for mapping and cataloging locations of invasive species within the State of Delaware.

- The DNREC conducts surveying and mapping of aquatic vegetation in the ponds to determine invasive and rare species.

A.9.6. RESEARCH NEEDS

- Maps of areas with high populations of invasive species are needed.

- A watershed approach needs to be taken when working with landowners in order to better prevent invasive species spread.

- A database of effective control methods for invasive species is needed.

A.10. SUMMARY OF AQUATIC INVASIVE SPECIES MANAGEMENT IN FLORIDA

A.10.1. AQUATIC INVASIVE SPECIES MANAGEMENT PLAN

Florida published its Statewide Invasive Species Strategic Plan for Florida in 2003.

A.10.2. AQUATIC INVASIVE SPECIES PROGRAMS AND ACTIVITIES

- **Aquatic Plant Management Program, Florida Department of Environmental Protection (FDEP), Bureau of Invasive Plant Management.** The program designs, funds, coordinates, and contracts invasive non-native aquatic plant control efforts in Florida's 1.25 million acres of public waters.

- **Annual survey for new infestations, FDEP.** Each year, 16 field biologists, each responsible for a particular region, conduct inventories in all 460 public water bodies (containing most of the state's surface water).

- **Hydrilla maintenance control, FDEP.**

- **Plant management services, FDEP.** Regional biologists are available to provide plant management services, such as consultation and guidance, to private and public landowners or managers.

- **Aquatic invasive species (AIS) control, South Florida Water Management District.** This Approach to control includes weekly treatment schedules and water use restrictions for aquatic herbicides.

- **Maintenance, Southwest Florida Water Management District (SWFWMD).** Aquatic plants including invasive species are controlled to maintain the flow capacity of flood control systems.

- **Surface Water Improvement Program, SWFWMD.** This restoration project is primarily geared towards preserving or restoring habitat and water quality. As part of restoring the natural hydrology of certain wetlands, the district plants a number of upland and aquatic native plants and also manages invasive plants in estuarine areas and lakes.

- **Mitigation Program, Florida Department of Transportation (FDOT) and SWFWMD.** FDOT funds a wetland mitigation program to compensate for road construction damage. The program involves preservation and restoration of native habitats, including invasive plant management and replanting of native vegetation.

- **Aquatic Plant Control (APC) Program, U.S. Army Corps, Jacksonville District.** This is a cost-share program with the state for control efforts in public water bodies.

- **Non-Native Fisheries Laboratory/Non-Native Fish Research Lab, Florida Fish and Wildlife Conservation Commission.** The Laboratory focuses its research on the 34 non-native fish species that have reproducing populations in Florida.

A.10.3. CLIMATE CHANGE CONCERNS

- Climate change may cause more hurricanes, which decreases the likelihood for hydrilla to grow. Significant amounts of rain and floodwater hinder the growth of hydrilla due to the resulting reduction in sunlight.

- Many AIS live in southern Florida, and climate change could result in northward range expansions. If the species make it to the northern part of the peninsula, they eventually could move into other states.

A.10.4. CLIMATE CHANGE ACTIONS

- An annual inventory may allow the state to observe and understand changes in invasive species populations over time.

A.10.5. RESEARCH ACTIVITIES AND INFORMATION USED

- As of January 2003, the Non-Native Fish Research Lab is responsible for assessing the role of 32 exotic fishes with reproducing populations in Florida. These fish include: the introduced walking catfish and swamp eel from Southeast Asia, tilapia from Africa, the Mayan cichlid from Central America, and the butterfly peacock from South America.

A.10.6. RESEARCH NEEDS

- Information that could improve the efficacy of herbicides and the timing of herbicide treatments is needed.

- Further investigation of selectivity issues is needed. The SWFWMD tries to be as selective as possible in targeting invasive plants and protecting/promoting the recovery of native plant communities by adjusting the timing of herbicide treatments, application rates, and treatment techniques in order to maximize treatment selectivity.

A.11. SUMMARY OF AQUATIC INVASIVE SPECIES MANAGEMENT IN GEORGIA

A.11.1. AQUATIC INVASIVE SPECIES MANAGEMENT PLAN

Plan under development.

A.11.2. AQUATIC INVASIVE SPECIES PROGRAMS AND ACTIVITIES

- **Aquatic Invasive Species (AIS) management activities, Georgia Department of Natural Resources (GADNR), Wildlife Resources Division, Fisheries Management.** The department responds to problematic invasive species with monitoring, containment, and removal. Giant salvinia, a primary problem, is being controlled with chemical treatments. Applesnail control and management includes surveys, destroying egg masses, and an applesnail task force was scheduled to be initiated in December 2005.

- **Swamp eels management, GADNR, Wildlife Resources Division, Fisheries Management; University of Georgia; U.S. Fish and Wildlife Service; and National Park Service.** Officials have been periodically surveying for the eel since its discovery in the late 1980s in artificial ponds at a nature center. The surveys in these ponds have occurred once a month since 2004. The next step will be to develop control recommendations.

- **Flathead catfish control program.** In 2006 the Georgia legislature allocated funding to control and manage invasive flathead catfish in Georgia. The increase in funding allowed for a fishery's biologist and two fishery's technicians to be hired to work on eradication and control methodologies.

- **Survey of lakes and reservoirs, Georgia Power (a regional utility), GADNR, Wildlife Resources Division, Fisheries Management.** Georgia Power surveys its lakes and reservoirs three or four times a year for aquatic invasive plants and applies spot treatments of herbicides when they are found. GADNR assists with these activities.

A.11.3. CLIMATE CHANGE CONCERNS

- Climate change is a potential threat to applesnail control efforts. If climate change results in warmer temperature at higher latitudes, the snail may have the potential to expand its habitat.

A.11.4. CLIMATE CHANGE ACTIONS

(None reported.)

A.11.5. RESEARCH ACTIVITIES AND INFORMATION USED

(None reported.)

A.11.6. RESEARCH NEEDS

- Interstate communication to prevent travel across borders with illegal exotic species is needed.

- Access to taxonomists is needed to correctly identify and learn about species.[1]

- Official state program on invasive species that includes a systematic control approach and organized response needs to be developed.

- Better ways of communicating with the public about invasive species needs to be developed.

- More information about control and capture methods and the ecological impacts of invasive species needs to be collected.

- More herbicide options and ways to expedite the registration process for new herbicides need to explored.

- An investigation of the human dimension of AIS introductions (i.e., intentional actions such as aquarium dumping, stocking or relocation and unintentional actions such as escapes of water garden species or use of invasives in landscaping and ornamental projects) needs to be conducted.

- The effectiveness of AIS outreach and education efforts as a means of modifying behavior (i.e., decreased releases and increased reporting) needs to be evaluated.

- Ecological and economic impacts of invasive non-native aquatic plant species in Georgia needs to be evaluated.

- Biological or alternative control methods for flathead catfish in south Georgia need to be identified.

- Efficacy of containment, control, or eradication activities for Asian swamp eels is needed.

- Early detection and surveillance plans coupled with response protocols should be developed.

- Database and GIS systems need to be developed with emphasis on interagency/interstate data sharing and user-friendly public access or report generating capabilities.

- Efficacy of channeled applesnail (CAS) control methods and techniques needs to be improved.

---

[1] Since the research was completed for this analysis, the need for a database of taxonomists has been fulfilled (see *ANS Task Force Experts Database* at http://www.anstaskforce.gov/experts/search.php).

- CAS risk assessment and thermal and salinity tolerance studies need to be conducted.

- Tilapia risk assessments and temperature and salinity tolerance research pertaining to culture activities need to be conducted.

A.12. SUMMARY OF AQUATIC INVASIVE SPECIES MANAGEMENT IN HAWAII

A.12.1. AQUATIC INVASIVE SPECIES MANAGEMENT PLAN

Hawaii completed its Aquatic Invasive Species (AIS) Management Plan in 2003 (see Appendix B, State Aquatic Invasive Species Management Plan Summaries for a general description of the Plan).

A.12.2. AQUATIC INVASIVE SPECIES PROGRAMS AND ACTIVITIES

- **AIS Response Team, Hawaii Department of Land and Natural Resources (DLNR), Division of Aquatic Resources (DAR).** The program team conducts the following activities (often in partnership with other agencies, universities, and organizations): surveying Lake Wilson for *Salvinia molesta*; controllingGorilla Ogo Algae, snowflake coral, and *Actinodiscus sp.*; mapping the distribution of invasive algae statewide; participating in hull fouling surveys of vessels traveling to the Northwest Hawaiian Islands Marine Sanctuary.

- **Coordinating Group for Alien Pest Species, multi-agency partnership.** This coordinating body facilitates communication among agencies, conducts public outreach, and increases awareness through various media campaigns. Eight public clean-up events have been coordinated to manually remove the invasive algae *Gracilaria salicornia* on Waikiki Beach.

- **Invasive Species Committees (ISCs) for island-based rapid response.** The ISCs are voluntary partnerships of private groups, government agencies, non-profit organizations, and concerned individuals working to protect each island from the negative impacts caused by invasive species. The overall goal of the ISCs is to prevent, eradicate, or control priority incipient invasive plant and animal species that threaten Hawaii's most intact federal, state, and private conservation lands. ISCs are are concerned with terrestrial alien species almost exclusively and are not involved in most AIS programs.

- **Plant Quarantine Branch, Department of Agriculture, Plant Industry Division.** This Division works with community groups that help to police the Central Oahu Lake by manually removing plants or by spot spraying using Aquamaster.

- **Hawaiian Ecosystems at Risk, Hawaii Cooperative Studies Unit, U.S. Geological Survey (USGS).** This project provides Internet technology, methods, and information to decision makers, resource managers, and the general public to help support effective, science-based management of harmful non-native species in Hawaii and the Pacific. Currently this project is funded by the National Biological Information Infrastructure/Pacific Basin Information Node through USGS/Pacific Islands Ecosystem Research Center.

- **AIS Advisory Group, DLNR, DAR.** This council sets priorities in AIS management. It is composed of members of federal, state, and other organizations involved in AIS issues.

A.12.3. CLIMATE CHANGE CONCERNS

- Climate change is linked to the increase in mosquito populations (which have an aquatic life stage), which reduces the population of local forest birds.

- Increased levels of greenhouse gases may negatively impact corals. A recent study conducted by a coral reef biologist from the Hawaii Institute of Marine Biology found that coral does not produce as much calcium carbonate under increased levels of carbon dioxide, because higher levels of carbon dioxide can result in a decline in ocean pH, leading to more acidic conditions.

A.12.4. CLIMATE CHANGE ACTIONS
(None reported.)

A.12.5. RESEARCH ACTIVITIES AND INFORMATION USED
(None reported.)

A.12.6. RESEARCH NEEDS

- Implementation of effective quarantine methods for incoming organisms is needed.

- Hawaii AIS management would benefit from more efficient detection methods for the newest invasive species. There is also a need for a better understanding of species range, including whether or not ranges are expanding. While officials have GIS capabilities, a staff shortage limits the number of people available to update range maps and do the field work.

- Information about how to smother the mushroom anemone is needed.

- Officials are developing a proposal for a literature review and research on effective control chemicals that will not harm coral reefs.

- There is a need for mechanisms to predict incoming invasive species.

- There is a need for more information on control methods, including biocontrols.

- Technology on cleaning hulls easily and safely is needed.

- There is a need for information and technology for the control of aquaculture releases. While the supersucker is being tested on algae, it is not practical for all areas, especially for shallower reefs.

- There is a need for chemical control methods for applesnails, which escaped from aquaculture ponds and invaded taro wetlands. The use of copper is too damaging.

- More effective control methods for giant reed are needed. Glyphosate is not effective enough. Arsenal™, a type of herbicide, is another option, but officials are unsure if it can be used in water. They need to know more about the non-target effects. Giant reed is harder to kill than many plants because of the depth of its root system. Another problem is locating existing populations. A developing method of thermal location would be very helpful, but it is still in the developmental stage.

- There is a need for better techniques for surveillance and detection. Officials rely strongly on the general public to report unusual events. Hiking groups and fishermen report such events often, but, without this information, Hawaii would have no way to know what is happening. There are not enough staff to carry out surveillance.

- Mechanisms to keep aquarium releases from occurring are needed.

- There is a need for *Salvinia molesta*, *Pistia*, and *Eichhornia* control and prevention.

A.13. SUMMARY OF AQUATIC INVASIVE SPECIES MANAGEMENT IN IDAHO

A.13.1. AQUATIC INVASIVE SPECIES MANAGEMENT PLAN
Plan under development.

A.13.2. AQUATIC INVASIVE SPECIES PROGRAMS AND ACTIVITIES

- **Noxious Weeds Program, Idaho State Department of Agriculture (ISDA), Cooperative Weed Management Area (CWMA).** CWMA is a distinguishable hydrologic, vegetative, or geographic zone based upon geography, weed infestations, and climatic or human-use patterns. CWMAs are formed when the landowners and land managers of a given area come together and agree to work cooperatively to control weeds. Idaho has 40 CWMAs, which are part of the ISDA cost-share program (the majority of the funding comes from federal sources). There are currently several cost-share participants that are working to deal with aquatic species in Idaho. The ISDA is responsible for administering the CWMA program in Idaho.

- **Eurasian Watermilfoil (EWM) Control Program, ISDA.** In response to the continuing economic and environmental crises created by EWM in Idaho's waters, the Idaho State Legislature appropriated $4 million to the ISDA for eradication and control of the aquatic weed. The Legislature directed these funds to be expended over a 2-year period beginning July 1, 2006 and ending June 30, 2008.

- **Invasive Species Council, Governor's Office.** The Idaho Invasive Species Council was established by Idaho's Governor's Executive Order in 2001. This Council is carrying out an inventory of EWM and conducting a public awareness campaign for boaters. The Invasive Species Coordinator is housed within ISDA.

- **EWM Task Force, Invasive Species Council.** The EWM Task Force was formed in 2002 to assist the Invasive Species Council in surveys and other EWM activities. The major activities of the Task Force include: (1) physically surveying all waters in the state; (2) developing a survey for all counties to prioritize actions and activities based on susceptibility factors; (3) engaging in multiple research projects with the University of Idaho (including research on control technologies); and (4) researching different herbicide combinations and exploring the use of new products.

- **Purple Loosestrife Control Efforts, University of Idaho.** This program uses biocontrol for purple loosestrife. Also, outreach programs both distribute insects (~40,000 distributed) and educate land managers on how to use them.

A.13.3. CLIMATE CHANGE CONCERNS

- In conducting the initial assessment for the Aquatic Invasive Species (AIS) Action Plan, officials considered latitude, longitude, temperature bands, elevation, and rainfall.

A.13.4. CLIMATE CHANGE ACTIONS

(None reported.)

A.13.5. RESEARCH ACTIVITIES AND INFORMATION USED

- Idaho is carrying out genetic analysis of *Myriophyllum* species and potential hybrids within the state.

- Physical surveys and mapping are being conducted

- Surveys of the Northern Idaho Lake are being done to determine the densest areas of EWM. These areas are then targeted with appropriate control methods. The less-dense areas are targeted by divers using hand-pulling techniques (removal of plants by the roots followed by vacuuming).

A.13.6. RESEARCH NEEDS

- Information on biocontrol methods is needed. This may require visits to the country of origin to examine the species under consideration.

- Reliable and continual funding is needed.

- Effective controls based on population size and the presence of other species are needed.

- Information on both the economic and ecosystem-related effects of specific AIS is needed.

- An effective herbicide, with less environmental impact, that can be applied in smaller amounts is needed. Researchers are currently looking for this type of herbicide.

- Bottom barriers are needed. Researchers are assessing the duration of placement for effective control and the potential for growth of aquatic plants after sediments have settled on the barriers.

- A soil-mix company that will recycle the milfoil into a soil mix is needed.

- Better ways need to be identified for more state partners to educate the public about why it is important to control EWM. A national or statewide database that could provide up-to-date information on current research being done for each invasive species would also be helpful.

A.14. SUMMARY OF AQUATIC INVASIVE SPECIES MANAGEMENT IN ILLINOIS

A.14.1. AQUATIC INVASIVE SPECIES MANAGEMENT PLAN

Illinois's State Comprehensive Management Plan for Aquatic Nuisance Species (ANS) was published in 1999 (see Appendix B, State Aquatic Invasive Species Management Plan Summaries for a general description of the Plan).

A.14.2. AQUATIC INVASIVE SPECIES PROGRAMS AND ACTIVITIES

- **Permanent Electric Dispersal Barrier, Illinois Department of Natural Resources (IDNR).** This barrier was initially designed to stop the round goby but is now being used to target other fish such as the big head silver carp.

- **Evaluation of Barriers, IDNR, Illinois Natural History Survey (INHS).** This program involves an evaluation of barriers to prevent the spread of bighead carp into the Great Lakes. It has conducted an assessment of multiple barrier components, including sonic technology, bubble arrays, and hydro-acoustic generators. INHS is also conducting field monitoring for the potential impacts of steel-hulled barges on movement of fish across an electric barrier to prevent entry of invasive carp into Lake Michigan.

- **Field Assessment of Electric Barrier in Chicago Sanitary and Ship Canal, IDNR-Fisheries.** IDNR is conducting monitoring of the existing electric demonstration barrier, including stocking and subsequent monitoring of radio and acoustic tagged fish (common carp) near the electric dispersal barrier in order to determine if they can move back and forth across the barrier.

- **Early Detection and Rapid Response Planning, IDNR, Fisheries.** IDNR is developing rapid response strategies for the control of Asian carp in various situations at critical control points and has educated their biologists and law enforcement officers on identifying various aquatic invasive species (AIS). If they find a species that is either new to the Illinois or new to a particular waterway/area, they are to fill out a standardized form and report it. This option is also available for the public in northern Illinois to track the Asian carp. If a species is reported, detection is verified, then IDNR follows up with a rapid assessment.

- **Bighead Carp Competition Studies, IDNR, INHS.** This program involves field monitoring and includes examining bighead carp competition with native filter feeding fish to assess the potential threat for Great Lakes fish (salmon and trout). IDNR is also examining bighead carp feeding on alewife and gizzard shad (food sources for salmon and trout).

- **Intensified Field Monitoring for Asian carp, IDNR, Fisheries.** This intensified field monitoring examines bighead and silver carp near Lockport and Brandon Road Pools at confluence of Des Plaines River.

- **Upper Illinois River habitat mapping, IDNR, INHS.** INHS is conducting field monitoring to evaluate Asian carp habitat.

- **Technical assistance for market development, IDNR, Fisheries, Illinois Department of Commerce and Economic Opportunity (DCEO).** The harvest program provides technical assistance for required analytical data to establish markets for Asian carp. Illinois's DCEO has provided implementation costs for start-up and phase 1 of an intensified harvesting program.

- **Contaminant analysis for market development, IDNR, INHS, University of Illinois.** The program provides additional contaminant analysis for market development.

- **Goby round-up/Carp Corral, a joint program with IDNR, U.S. Army Corps of Engineers, U.S. Fish and Wildlife Service, the Metropolitan Water Reclamation District.** The program monitors the spread and expansion of round goby and bighead/silver carp populations in the Illinois River System toward Lake Michigan.

- **Eradication, IDNR.** IDNR conducts eradication of Eurasian watermilfoil using Chemical- 2-4D and sonar. The agency is also experimenting with treatment timing and dosage for better long-term effects. A new project will target curlyleaf pondweed.

- **Permanent Electric Dispersal Barrier, IDNR.** An electric barrier has been implemented in the Chicago Sanitary and Ship Canal to deter the inter-basin transfer of invasive fish between the Great Lakes and Mississippi River. The barrier will be operated and funded by the IDNR upon completion; in the interim, the U.S. Army Corps of Engineers maintains management of the barrier.

A.14.3. CLIMATE CHANGE CONCERNS

- Climate change may have an indirect impact by allowing some species to expand into new ranges where they have not historically been found. If certain regions warm up (or cool down), they may be colonized by species that were only marginally adapted to the warmer (or cooler) temperatures.

- Illinois's ANS Plan includes vectors that are exacerbated by climate change:

    As use of the Great Lakes intensified as a transport route for commerce, the rate of introduction of aquatic nuisance species also increased. More than one-third of the organisms have been introduced in the last 30 years, a surge coinciding with the opening of the St. Lawrence Seaway. Other human activities contributing to the transport and dispersal of aquatic nuisance species in the Great Lakes and inland state waters include the release of organisms from the ballast water of ships, transport and release from the bottoms of ships, movement or intentional release of aquaculture and sport fishery species along with their associated (free living and parasitic) organisms, release of organisms associated with pet industries or pest management practices, recreational boating, bait handling, water transport, and ornamental and landscape practices.

See Illinois State Comprehensive Management Plan for Aquatic Nuisance Species (1999).

A.14.4. CLIMATE CHANGE ACTIONS

(None reported.)

A.14.5. RESEARCH ACTIVITIES AND INFORMATION USED

(None reported.)

A.14.6. RESEARCH NEEDS

- Research on Asian carp is needed. IDNR needs a good understanding of their specific reproduction requirements, biomass and population estimates, preferred habitats, and the effects of competition with Great Lakes native fish. Officials would like to know how many invasive fish exist, their size, and where they are located, in order to better target them.

- Examination of the consistencies and inconsistencies between different state laws is needed. Many state laws are changing and, if the National AIS Act is passed, it will be important to know what the states are all doing in this area.

A.15. SUMMARY OF AQUATIC INVASIVE SPECIES MANAGEMENT IN INDIANA

A.15.1. AQUATIC INVASIVE SPECIES MANAGEMENT PLAN

Indiana's Aquatic Nuisance Species Management Plan was published in 2004 (see Appendix B, State Aquatic Invasive Species Management Plan Summaries for a general description of the Plan).

A.15.2. AQUATIC INVASIVE SPECIES PROGRAMS AND ACTIVITIES

- **Early Detection and Rapid Response, Indiana Department of Natural Resources, (INDNR), Division of Fish and Wildlife.** The Division is conducting treatment of Brazilian elodea in Griffy Lake, as well as a survey and development of an aquatic vegetation management plan. Whole-lake herbicide treatment began in 2006 and will continue in 2007. Access restrictions were implemented in the spring of 2006 to prevent the movement of Brazilian elodea to other waters. The Division is also conducting rapid response for hydrilla, first discovered in 2006 at Lake Manitou. Response included an herbicide treatment and access restrictions in the fall of 2006. Large-scale aquatic herbicide treatments are planned for the spring of 2007.

- **Lake and River Enhancement Program, INDNR Division of Fish and Wildlife.** The Program provides grants to lake associations for the control of aquatic invasive plants.

- **Yellow Perch Research, INDNR Division of Fish and Wildlife, Ball State University.** Researchers examine the impacts of aquatic invasive species (AIS) such as zebra mussels and round goby on yellow perch and other native species in Lake Michigan.

- **Management of sport fisheries, INDNR Division of Fish and Wildlife.** The Division is responding to AIS' threats to sport fisheries through the following actions: (1) eradicating fishery altogether; (2) stocking predators; and (3) manipulating habitat (e.g., lake drawdowns to reduce aquatic invasive fish and plants).

- **Emergent aquatic plant control, INDNR Division of Nature Preserves.** The Division is controlling purple loosestrife, Phragmites, and reed canary grass on Indiana's nature preserves. A purple loosestrife biocontrol program has been implemented using beetles on Nature Preserve properties as well as other areas that contain large areas of purple loosestrife. Phragmites and reed canary grass have been sprayed with glyphosate-based herbicides, though the Division uses some plant-specific herbicides for reed canary grass. The Division has also performed some herbicide control for the narrowleaf cattail and hybrid cattail, though there is some debate about whether narrowleaf cattail is native to North America.

A.15.3. CLIMATE CHANGE CONCERNS

(None reported.)

A.15.4. CLIMATE CHANGE ACTIONS

(None reported.)

A.15.5. RESEARCH ACTIVITIES AND INFORMATION USED

(None reported.)

A.15.6. RESEARCH NEEDS

- Development of effective ballast water treatment technologies for the Great Lakes is needed.

- Research on biocontrol for Phragmites or reed canary grass needs to be conducted. There is an active research program to develop biocontrol for Phragmites at Cornell University, but more research should be devoted to developing herbicides that are highly selective for these plants to reduce damage to non-target wetland plants.

- Further investigation of biocontrols for Eurasian watermilfoil, curlyleaf pondweed, hydrilla, and Brazilian elodea is needed.

- Continued refinement of herbicides and timing of applications to reduce non-target plant damage is needed.

A.16. SUMMARY OF AQUATIC INVASIVE SPECIES MANAGEMENT IN IOWA

A.16.1. AQUATIC INVASIVE SPECIES MANAGEMENT PLAN

Iowa's Plan for the Management of Aquatic Nuisance Species was published in 1999 (see Appendix B, State Aquatic Invasive Species Management Plan Summaries for a general description of the Plan).

A.16.2. AQUATIC INVASIVE SPECIES PROGRAMS AND ACTIVITIES

- **Iowa Invasive Species Working Group.** This group of federal, state, county, and University staff hold regular meetings to discuss invasive species issues and plans.

A.16.3. CLIMATE CHANGE CONCERNS
(None reported.)

A.16.4. CLIMATE CHANGE ACTIONS
(None reported.)

A.16.5. RESEARCH ACTIVITIES AND INFORMATION USED
(None reported.)

A.16.6. RESEARCH NEEDS

- More information on the ways to control zebra mussels in the environment is needed.

- More information on Asian carp, including control methods, biological information about the species, and documentation of their impacts needs to be collected.

- Information on aquatic invasive species not yet found in northern climates that are capable of surviving in colder climates or that may expand their ranges due to climate changes needs to be gathered.

- Faster rapid-response systems need to be developed and funding mechanisms to implement them need to be identified.

- Increased public awareness about invasive species is needed.

A.17. SUMMARY OF AQUATIC INVASIVE SPECIES MANAGEMENT IN KANSAS

A.17.1. AQUATIC INVASIVE SPECIES MANAGEMENT PLAN

Kansas completed its Aquatic Nuisance Species (ANS) Management Plan in April 2005 (see Appendix B, State Aquatic Invasive Species Management Plan Summaries for a general description of the Plan).

A.17.2. AQUATIC INVASIVE SPECIES PROGRAMS AND ACTIVITIES

- **Plant Protection and Weed Control Program, Kansas Department of Agriculture.** The agency has regulatory authority to deal with aquatic invasive weeds and conducts quarantines on purple loosestrife, tamarisk, and all federal noxious weeds, including the 19 aquatic species.

- **ANS Program, Kansas Department of Wildlife and Parks.** The program is designed to protect residents of Kansas and aquatic resources from the effects of ANS. The program focuses on preventing the accidental introduction of new ANS, limiting the spread of existing aquatic invasive species (AIS), and controlling or eradicating AIS where environmentally and economically feasible. The intentional introduction of non-indigenous species for aquaculture, commercial, or recreational purposes is managed to insure that these beneficial introductions do not result in accidental AIS introductions. The Program also seeks to improve information sharing among those agencies responsible for regulation of intentional introductions.

A.17.3. CLIMATE CHANGE CONCERNS

- New AIS threats to Kansas's aquatic resources may emerge as a result of a shift in the climate.

A.17.4. CLIMATE CHANGE ACTIONS

- Kansas State officials consider climate change by communicating with colleagues to the north and south about species that are moving into the state and by comparing response activities.

A.17.5. RESEARCH ACTIVITIES AND INFORMATION USED

- Boater movement surveys are being implemented.

- Risk assessments are being conducted.

- Research on zebra mussels is being carried, specifically on movement via live-wells and bilges (veliger stage) and population dynamics.

A.17.6. RESEARCH NEEDS

- Research on the effects of AIS on water quality is needed.
- Research on Asian tapeworm presence is needed.
- Research on zebra mussel eradication techniques needs to be conducted.
- AIS vectors and exclusion techniques need to be identified.
- Research on effective public outreach tools and rapid response is needed.

A.18. SUMMARY OF AQUATIC INVASIVE SPECIES MANAGEMENT IN KENTUCKY

A.18.1. AQUATIC INVASIVE SPECIES MANAGEMENT PLAN
Plan under development.

A.18.2. AQUATIC INVASIVE SPECIES PROGRAMS AND ACTIVITIES

- **Integrated Roadside Vegetation Management Program, Kentucky Department of Highway, Roadside Branch.** Through herbicides and mowing, the program controls noxious weed species along highway rights-of-way. Fertilization is also conducted to encourage rapid root growth of other plants.

- **Control Program, Kentucky State Nature Preserves Commission, The Nature Conservancy and Northern Kentucky University.** KNSPC works to systematically control and contain invasive plants within the nature preserve system statewide. Control mechanisms include cutting and removal as well as herbicide applications. Fire is also being tested as a tool to control the plants.

- **Control Program, University of Kentucky/Lexington-Fayette Urban County Government (grant funds from the Columbus Advisory Board).** The program removes invasive plants from Arboretum Park.

- **Control program, Kentucky Department of Fish and Wildlife Resources (KDFWR).** The Department controls populations of big head and silver carp by allowing commercial harvesting of the fish.

- **Monitoring and research program, KDFWR.** The Department is conducting research on cormorants to understand how they live, what they eat, and the impacts they have on habitats.

- **University of Kentucky Invasive Species Initiative.** The program, initiated in 2006, is using an interdisciplinary approach to monitor, model, prevent, mitigate, and eradicate aquatic and terrestrial invasive species in Kentucky.

- **Tracy Farmer Center for the Environment at University of Kentucky.** Using a hands-on approach, this youth outreach program teaches students about invasive species. They work to incorporate invasive species awareness into secondary school science curriculums across the state.

A.18.3. CLIMATE CHANGE CONCERNS
(None reported.)

A.18.4. CLIMATE CHANGE ACTIONS

(None reported.)

A.18.5. RESEARCH ACTIVITIES AND INFORMATION USED

- Research on how to limit fish populations, including bighead and silver carp is being conducted.

A.18.6. RESEARCH NEEDS

- More information on the commercial value and uses of big head and silver carp is needed.

- General research on the cormorant is needed.

- A Kentucky aquatic biodiversity database needs to be developed to track distribution of aquatic organisms (native and invasive) across the state.

- An assessment of aquatic invasive species impacts on endangered and threatened flora and fauna (especially mussels) and on fisheries is needed.

- Assessment of potential biocontrols on native flora and fauna is needed.

A.19. SUMMARY OF AQUATIC INVASIVE SPECIES MANAGEMENT IN LOUISIANA

A.19.1. AQUATIC INVASIVE SPECIES MANAGEMENT PLAN

Louisiana's State Management Plan for Aquatic Invasive Species (AIS) was published in July 2005 (see Appendix B, State Aquatic Invasive Species Management Plan Summaries for a general description of the Plan).

A.19.2. AQUATIC INVASIVE SPECIES PROGRAMS AND ACTIVITIES

- **The Louisiana Aquatic Invasive Species Task Force, chaired by the Louisiana Department of Wildlife and Fisheries (LDWF) and composed of state and federal agencies, stakeholders, and industry groups.** The Task Force completed a draft AIS plan in 2005 and advises the Louisiana AIS Council, a permanent working partnership charged with implementation of the state AIS management plan.

- **Aquatic Plant Control Fund.** The fund was created by the state legislature for the control of nuisance aquatic vegetation. At present, the fund is derived solely from an increase in boat trailer registration fees.

- **Aquatic Plant Management Program, LDWF.** This program maintains boating and fishing access through herbicide applications to nuisance aquatic vegetation.

- **Aquatic Animals Management Program, LDWF.** LDWF has posted a bounty on the tails of nutria. The goal is to obtain 600,000 tails per year. The Department is also monitoring to see if marshes are recovering.

- **Outreach activities, conducted by many organizations that use some state funds in addition to other funds, including Louisiana Sea Grant College Program, Barataria-Terrebonne Estuary Program, and The Nature Conservancy, among others.** Outreach is focused on target audiences (i.e., recreational fishers, water gardeners, and aquaculture groups) and elementary school children.

A.19.3. CLIMATE CHANGE CONCERNS

- Climate change will make conditions more suitable for some species and less suitable for other species.

- The state may experience land loss due to rising sea levels.

A.19.4. CLIMATE CHANGE ACTIONS

(None reported.)

A.19.5. RESEARCH ACTIVITIES AND INFORMATION USED

(None reported.)

A.19.6. RESEARCH NEEDS

- Satellite technology to determine the location of invasive species is needed.

A.20. SUMMARY OF AQUATIC INVASIVE SPECIES MANAGEMENT IN MAINE

A.20.1. AQUATIC INVASIVE SPECIES MANAGEMENT PLAN

Maine's Action Plan for Managing Invasive Aquatic Species: A Report to the Land and Water Resources Council from the Interagency Task Force on Invasive Aquatic Plants and Nuisance Species was published in October 2002 (see Appendix B, State Aquatic Invasive Species Management Plan Summaries for a general description of the Plan).

A.20.2. AQUATIC INVASIVE SPECIES PROGRAMS AND ACTIVITIES

- **Invasive Aquatic Plant Prevention Program, Maine Department of Environmental Protection (ME DEP).** This program inspects watercraft, trailers, and outboard motors at or near the state borders and at boat launching sites for the presence of invasive aquatic plants. The program also provides educational materials to the public and to watercraft owners on invasive aquatic plants and funds control work by some private lake associations. The Program is also conducting plant control work on three small lakes (one with populations of Eurasian watermilfoil, one with hydrilla, and one with curlyleaf pondweed) to try to prevent the spread of these plants to other water bodies. Finally, the agency is also undertaking plant removal on lakes with variable milfoil located close to boat ramps in order to reduce spread.

- **Invasive Aquatic Plant Prevention Program, Maine Department of Inland Fisheries and Wildlife (MDIFW).** MDIFW has a warden service to patrol waters and roads and enforce violations like launching a boat or transporting a vehicle on public roads with plants attached.

- **Lake and River Protection Sticker, ME DEP, MDIFW.** As of 2002, all motorized watercraft on inland waters in Maine are required to display the Lake and River Protection Sticker ("Preserve Maine Waters"). No sticker is required for operating a boat in tidal waters. Motorized watercraft includes any boat with any type of motor, including canoes with electric motors and personal watercraft. Dedicated funds raised through this program are used to support Maine's prevention and early detection and rapid response efforts. Maine raises approximately $1 million a year through this program.

- **Courtesy Boat Inspection Program, ME DEP.** The program involves voluntary boat inspections that emphasize boat ramp inspection. Last year, there were 30,000 inspections (10,000 more than in 2003).

- **Early Detection, Invasive Plant Patrol Program, ME DEP.** ME DEP contracts with the Volunteer Lake Monitoring Program, which, through the Maine Center for Invasive Aquatic Plants (MCIAP), conducts training programs for volunteers, state agency personnel, professionals, teachers, students, and others. Since the program began in 2001, nearly 1,400 individuals have been trained. The basic workshop teaches participants how to recognize the invasive plants on Maine's "eleven most unwanted" list and how to distinguish these invaders from the native species they resemble. A variety of advanced training opportunities is also offered. The number of Maine water bodies being screened for the presence of invasive aquatic plants has increased several hundredfold since MCIAP began its training effort. Surveys conducted by volunteers now account for more than half of all surveys being conducted in the state. Maine inspects watercraft, trailers, and outboard motors and provides educational materials to the public. In order to decide which ramps to target, ME DEP conducts a rough risk assessment to determine which ramps are used most often. The Department uses paid inspectors for the high-use hours, to keep any invasive plants from spreading. Officials have completed a vulnerability assessment (remotely, using GIS) to assist the analysis, examining the distance from infested water bodies to highways and whether infested water bodies are hydrologically connected to other water bodies.

- **Draft Rapid Response Plan, ME DEP, MDIFW.** The Commissioners of the ME DEP and the MDIFW have agreed to direct their respective agencies' responses to new infestations of invasive aquatic species under the auspices of a single, coordinated rapid response plan. Species covered by the 172-page plan include invasive plants and fish already in some Maine waters and other exotic organisms not yet established in Maine, such as zebra mussels.

- **Integrated Pest Management Strategy (for purple loosestrife), Maine Department of Agriculture.** This program works to avoid water drawdown and site disturbance during the growing season to avoid exposing mudflats where seeds can germinate. The program surveys all wetlands (at least every 3 years) to pinpoint infestations, and every year, stems at "active" wetland sites are sprayed with the herbicide glyphosate and counted at selected sampling sites. Park authorities are beginning to work with landowners on sites adjacent to park boundaries to enact similar preventative strategies.

A.20.3. CLIMATE CHANGE CONCERNS

- Maine's Aquatic Invasive Species (AIS) Management Plan states that "…with global climate change, [AIS] may spread even further as freshwater and ocean temperatures moderate."

A.20.4. CLIMATE CHANGE ACTIONS

- The AIS Management Plan has a category entitled "No Action at This Time" that emphasizes the need to "[l]earn more before acting" (p. 14). The category lists climate change as an issue. Specifically, the Plan states that:

> Maine's cold climate and ocean temperatures now limit warm water species. But warming temperatures and fluctuating weather patterns may in time be more favorable to their introduction. At the same time, changing conditions may become less favorable for coldwater species, thus contributing to an overall shift toward

warm water assemblages. Taking the long view, Maine will monitor climatic conditions to provide early warning of potential infestations.

A.20.5. RESEARCH ACTIVITIES AND INFORMATION USED

- The state has recently completed a 2-year research project studying the relative effectiveness of different manual methods for controlling variable watermilfoil, as well as the viability of variable milfoil fragments under different conditions. This research will be continued in the future and will focus on the impacts of variable watermilfoil on native ecosystems.

- Students of one professor at the University of Maine at Farmington routinely are involved in invasive aquatic plant surveys, assessments, and mapping projects in Maine, as well as research on fragment regeneration.

A.20.6. RESEARCH NEEDS

- Research need to be conducted to find a native organism that can function as a safe, effective biocontrol for variable watermilfoil.

A.21. SUMMARY OF AQUATIC INVASIVE SPECIES MANAGEMENT IN MARYLAND

A.21.1. AQUATIC INVASIVE SPECIES MANAGEMENT PLAN
No plan available.

A.21.2. AQUATIC INVASIVE SPECIES PROGRAMS AND ACTIVITIES

- **Maryland Marsh Restoration/Nutria Project Partnership, led by Maryland Department of Natural Resources (MD DNR) and U.S. Fish and Wildlife Service (USFWS) in partnership with 24 additional federal, state, and private organizations.** The project involves behavioral/population research, reproductive research, testing of trapping methods, population control strategies, and marsh restoration.

- **Mute Swan Management, MD DNR, Wildlife and Heritage Service.** MD DNR manages the mute swan population through (1) public outreach and education; (2) population management and resource protection (e.g., reducing recruitment by egg oiling, humane removal of adult swans, establishing Swan-Free Areas); (3) regulating the possession of mute swans; (4) relief of human safety and nuisance conflicts; and (5) population monitoring and research.

- **Zebra Mussel Prevention, MD DNR.** This program educates boaters and divers about zebra mussels. The goal is to prevent mussels from becoming established in the state.

- **Water Chestnut Harvesting, MD DNR, Division of Tidewater Ecosystem Assessment.** Water chestnut, recently rediscovered in the Upper Chesapeake Bay, is pulled by hand during Submerged Aquatic Vegetation surveys.

- **Snakehead Prevention, MD DNR, Fisheries Service.** The service seeks to prevent the spread of snakeheads by conducting the following activities: circulating posters that ask anglers to kill and report all snakeheads; compiling regional data (database is maintained by the Virginia Department of Game and Inland Fisheries [VDGIF]) for captures in the Potomoc River (these include MD DNR, VDGIF, and public captures); and annual monitoring that includes seine, electrofishing, and gillnet surveys.

- **Snakehead Control and Management Plan, USFWS.** The creation of the Snakehead Control and Management Plan is a collaborative effort among industry, non-governmental organizations (NGOs), state and federal agencies, and citizens. The goal is to create a management plan that identifies action items to guide agency activities and funding priorities, in addition to goals for industry, citizens, and NGOs. The plan will focus on control priorities for the Potomac/Northeast U.S. region, as well as general prevention, early detection and rapid response, research, and outreach/education priorities in other regions the snakehead could potentially invade.

- **Purple Loosestrife Control, MD DNR.** Maryland biologists will manually cull purple loosestrife if encountered in the field. The state has also used biocontrols for several years.

- **Cooperative Giant Hogweed Eradication, MD DNR, Maryland Department of Agriculture (MDDA).** MD DNR works to eradicate giant hogweed by using a combination of hand-pulling, herbicide application, burning, and bagging techniques each summer.

- **Plant Pest Survey and Detection, MDDA, Plant Protection and Weed Management Section.** No specific action.

- **Phragmites Control Cost-Share Program, MD DNR, private citizens.** This program supplies private landowners with herbicides for Phragmites control. MD DNR or MDDA can apply the herbicides and bill landowners, or the landowner can use a private applicator. Landowners incur any application costs.

- **Aquatic Weed Control with Herbicides, MDDA, Plant Protection and Weed Management Section.** MDDA staff consider timing, permitting, organism's effect on ecosystem, expense, and level of effort required for control in deciding which herbicides to use and when to use them.

- **Chinese Mitten Crab Investigation, MD DNR.** The MD DNR, in cooperation with the Smithsonian Environmental Research Center, the USFWS, and NOAA has established a joint effort to evaluate the status of the Chinese mitten crab in Maryland.

A.21.3. CLIMATE CHANGE CONCERNS

- Climate change may affect the nutria problem.

- A rise in sea level may place additional stress on marshes, which are highly sensitive to changes in water level. Marsh resources, if any remain, will migrate landward. Marsh loss is caused by a combination of nutria and sea level rise and subsidence of the general terrain in the area.

- Significant warming may result in habitat changes, causing species such as the Bulls-Eye Snakehead, in Florida, to become an issue in Maryland.

A.21.4. CLIMATE CHANGE ACTIONS

- The U.S. Army Corps of Engineers will use an ongoing nutria study to implement a 4-year marsh restoration project, potentially covering 150 acres of marsh in the Blackwater National Wildlife Refuge. The Corps is using sediment spraying to raise the level of the marsh, which helps to restore the marsh grass.

A.21.5.   RESEARCH ACTIVITIES AND INFORMATION USED

(None reported.)

A.21.6.   RESEARCH NEEDS

- Research on nutria pheromonal attractants and weaknesses in reproductive biology is needed.

- Research on zebra mussels and their control techniques in lakes and rivers is needed.

- Species-specific fish control techniques are needed.

- Innovative control techniques for snakeheads that would allow officials to apply a lethal control need to be developed.

- Information on chemicals that would either attract fish or exclude them from areas needs to be collected.

- A contained area to study snakeheads in order to develop innovative techniques to sample and control them is needed.

- Information on better Phragmites control methods, other than herbicides (e.g., biocontrol) needs to be collected.

A.22. SUMMARY OF AQUATIC INVASIVE SPECIES MANAGEMENT IN MASSACHUSETTS

A.22.1. AQUATIC INVASIVE SPECIES MANAGEMENT PLAN

Massachusetts's Aquatic Invasive Species Management Plan was published in 2002 (see Appendix B, State Aquatic Invasive Species Management Plan Summaries for a general description of the Plan).

A.22.2. AQUATIC INVASIVE SPECIES PROGRAMS AND ACTIVITIES

- **Boat Ramp Monitor Program, Massachusetts Department of Conservation and Recreation (MA DCR), Office of Water Resources (OWR), Lakes and Ponds Program (LPP).** Boat ramp monitors are positioned at lakes and ponds statewide to inspect boats and ensure that no plant fragments are attached to the boat, trailer, or gear. Boaters are given informational brochures and asked to participate in a voluntary boat inspection and complete a survey. LPP posts aquatic invasive species (AIS) posters in kiosks and metal reflective boat ramps at public access points to remind boaters to check their boats and trailers before entering or leaving a water body.

- **Weed Watchers Program, MA DCR, OWR, LPP.** LPP schedules weed watcher training for any interested lake groups or associations. This program teaches groups how to check key areas such as inlets, outlets, and shallow areas. The training also teaches volunteers how to eradicate species. This program is modeled after New Hampshire's weed watcher program.

- **Multi-lingual Education, MA DCR, OWR, LPP, Massachusetts Office of Coastal Zone Management (MOCZM).** Lead by MOCZM with participation from LPP, this program developed multi-lingual brochures to distribute to specific groups (e.g., participants in the seafood trade who are Chinese).

- **Rapid Response Protocols, MA DCR, OWR, LPP.** MA DCR paid a contractor to develop rapid response protocols for new and unknown AIS.

- **AIS Program, MOCZM.** Recent projects by MOCZM include developing resources for early detection and rapid response to new invasions in Massachusetts, developing a Website to provide a single outlet for AIS information and resources in the state, and developing a marine invasive species monitoring network. The monitoring network uses a standardized protocol and identification resources developed with funding by MOCZM. The Office partnered with Massachusetts Institute of Technology (MIT) Sea Grant to develop a centralized marine invasive species data management system, as well as Massachusetts Bays National Estuary Program in an effort to train citizens to monitor along the coast. MOCZM has also taken steps to establish memoranda of understanding with state agencies to coordinate management and has launched efforts to engage the seafood and pet-store industry.

- **Massachusetts Bays National Estuaries Program.** The Program coordinated the 2003 rapid assessment survey of non-native and native marine species of floating dock communities with MIT Sea Grant. Another rapid assessment is scheduled for summer 2007. This program has also sponsored research and developed publications related to AIS.

- **MIT Sea Grant Program.** This program is leading the development of a centralized marine invasive species data management system. The database includes information from many groups, including volunteer monitors and divers. MIT Sea Grant are also used to develop informational publications that help minimize new introductions through several vectors.

- **Water Chestnut Eradication, U.S. Fish and Wildlife Service, Silvio O. Conte National Fish and Wildlife Refuge, in partnership with a number of other groups.** This program's control component consists of mechanical harvesting and some herbicide application around the edges of the water body. Participants hand pull the plant at six sites including Holyoke, Hadley, East Hampton, South Hadley, as well as a few sites in Connecticut. The plant is almost completely eradicated from sites where manual culling has been employed for the past 4 years.

- **Giant Hogweed control, Massachusetts Department of Agricultural Resources, Division of Regulatory Services.**

A.22.3. CLIMATE CHANGE CONCERNS

- As the climate warms, certain plants that pose problems in the south could move into Massachusetts. For example, water hyacinth, a problem in Virginia and other southern states, and which is being sold in nurseries for water gardens, is not yet considered a problem in the state. If the climate warms up enough to allow water hyacinth to overwinter, it could be a threat. In contrast, water chestnut cannot be legally possessed and is not traded in the marketplace.

A.22.4. CLIMATE CHANGE ACTIONS
(None reported.)

A.22.5. RESEARCH ACTIVITIES AND INFORMATION USED
(None reported.)

A.22.6. RESEARCH NEEDS
(None reported.)

A.23. SUMMARY OF AQUATIC INVASIVE SPECIES MANAGEMENT IN MICHIGAN

A.23.1. AQUATIC INVASIVE SPECIES MANAGEMENT PLAN

Michigan's Aquatic Nuisance Species (ANS) Management Plan was published in 2002 (see Appendix B, State Aquatic Invasive Species Management Plan Summaries for a general description of the Plan).

A.23.2. AQUATIC INVASIVE SPECIES PROGRAMS AND ACTIVITIES

- **ANS Council, Michigan Department of Environmental Quality (MDEQ); Michigan Department of Natural Resources (MI DNR); Michigan Department of Agriculture (MDA); National Wildlife Federation; Michigan United Conservation Clubs; Michigan Education Association; Michigan State University, Department of Fisheries and Wildlife.** The Council implements the ANS Management Plan and does planning and strategy for member agencies and associations. As of 2006, council members are considering a rapid response plan. The Council monitors aquatic invasive species (AIS) and promotes control, but not eradication (Michigan does not spend money to eradicate AIS where it is impossible). The Council also focuses on measures to prevent further introductions and spread of ANS.

- **Invasive Species Advisory Council, MDEQ, Michigan Department of Transportation, MI DNR, MDA.** The Council is responsible for overseeing all management of invasive species in the state (aquatic and terrestrial).

- **Education and Outreach, MDEQ Office of the Great Lakes.** MDEQ officials conduct outreach on how to prevent the spread of ANS. The office also offers removal and control training for local governments, conservation groups, citizens, and associations and issues permits for the use of chemicals for ANS removal.

- **Status and Trends Surveys, MI DNR, Fisheries Division.** When habitat biologists encounter AIS during their annual fish Status and Trend Surveys, they kill and preserve it for later identification. Any recurrence is noted in the files.

- **Purple Loosestrife Program, Michigan State University, Michigan Sea Grant College Program.** This program introduces biocontrol agents (natural insect enemies) to existing purple loosestrife populations.

A.23.3. CLIMATE CHANGE CONCERNS

- MDEQ is concerned about ANS expansion as waters warm. Hydrilla and water lettuce are overwintering in northern areas.

A.23.4. CLIMATE CHANGE ACTIONS

- Officials are addressing the overwintering of hydrilla and water lettuce in northern areas with outreach and education efforts.

A.23.5. RESEARCH ACTIVITIES AND INFORMATION USED

- The 2002 ANS Management Plan includes the following: research on treatment of ballast water; surveys of purple loosestrife throughout Michigan; research on the practicality of pheromone use as a control for round goby; assessment of impacts of round gobies and collection of baseline data on ruffe; and testing for effects of zebra mussel on zoobenthos and the diet and growth of yellow perch.

A.23.6. RESEARCH NEEDS

- The 2002 ANS Management Plan includes the following: (1) *Preventing AIS introductions*—including monitoring, data for rapid response, probabilities for establishment, hot list of potential AIS, boater and angler survey regarding implementation methods; (2) *Controlling AIS*—including use of biocontrol, pesticides, physical control, social/political/economic acceptability of control, effectiveness and pathways; (3) *Researching and monitoring AIS impacts*—including potential invasive risks of genetically modified aquatic plants and fish to Michigan's aquatic ecosystems and to aquaculture and sport fishing; and (4) *Capacity-building*—for AIS data and quality scientific research by promoting data availability and collaboration among agencies, researchers, and industry.

- Research on the impacts of controls is needed, especially chemical controls. Michigan officials aim to conduct a long-term costs/benefits analysis and evaluate the environmental impacts of AIS. The officials are interested in whether long-term studies will show the weevil beetle to be an effective milfoil biocontrol, as well as the impacts of control methods on water quality and ecosystem stability.

A.24. SUMMARY OF AQUATIC INVASIVE SPECIES MANAGEMENT IN MINNESOTA

A.24.1. AQUATIC INVASIVE SPECIES MANAGEMENT PLAN

Plan under development.

A.24.2. AQUATIC INVASIVE SPECIES PROGRAMS AND ACTIVITIES

- **Invasive Species Program, Minnesota Department of Agriculture.** The Department issues an annual report that characterizes aquatic plants and wild animals, and the annual report provides fund allocation records for aquatic invasive species (AIS) activities through watercraft surcharges and a water recreation account. The Department also has educational requirements for terrestrial species. The three primary goals of the program are to (1) prevent introductions, (2) prevent spread, and (3) reduce impacts.

- **Eurasian Watermilfoil Management (EWM) Program, Minnesota Department of Natural Resources (MN DNR).** This program focuses on unintentional transport of milfoil on boat equipment and better cleaning of such equipment by performing four functions: (1) monitoring milfoil growth; (2) coordinating with government agencies, special purpose districts, and lakeshore associations to prevent spread; (3) coordinating with the University of Minnesota and other facilities to study the use of biocontrols and herbicides; and (4) the allocation of grants to potential partners working on lakes with public water access and funds research on biocontrol.

- **Purple Loosestrife Management Program, MN DNR.** This program seeks to reduce the environmental effects of purple loosestrife by integrating chemical and biocontrol and cooperating with local, state, and federal groups.

- **Watercraft Inspection Internships, MN DNR.** Between April and October, watercraft inspections are conducted at public water access sites on lakes and rivers infested with AIS.

A.24.3. CLIMATE CHANGE CONCERNS

- Warmer climate species, which cannot currently survive in Minnesota, will eventually be able to survive in the state.

- Warming could produce cooler and wetter springs, which would limit the growth of EWM. Historically, drought conditions caused the initial growth of watermilfoil.

- As temperatures warm over time, conifers may be replaced by oaks, followed by prairie grassland. However, with invasive species, such as buckthorn, oaks may not be able to compete, which will throw off the natural cycle.

A.24.4. CLIMATE CHANGE ACTIONS

(None reported.)

A.24.5. RESEARCH ACTIVITIES AND INFORMATION USED

- Technology to deter the spread of Asian carp is being implemented.
- Funding is being used to construct dispersal barriers for Asian carp.
- Weekly copper sulfate treatments to kill zebra mussels are being implemented.
- Public awareness and watercraft inspections are being conducted.
- Technical assistance for curlyleaf pondweed management projects is being offered.

A.24.6. RESEARCH NEEDS

- Habitat recovery issues after eradication of an invasive species need to be examined.

- A national framework or law on invasive species to deal with intrastate transportation, transportation on public roads, and interstate transpiration is needed. The Lacey Act is not sufficient.

- States should be sharing more information on risk assessments.

- Information on effective herbicide and biocontrol methods is needed.

- A more comprehensive noxious weed list and a list of injurious wildlife need to be developed, as well as a list of federal experts that states can contact if they have questions on a particular issue.

- More studies on the effects of EWM need to be conducted.

- More information on long-term impacts of invasive species on adjacent wildlife communities is needed.

A.25. SUMMARY OF AQUATIC INVASIVE SPECIES MANAGEMENT IN MISSISSIPPI

A.25.1. AQUATIC INVASIVE SPECIES MANAGEMENT PLAN

Plan under development.

A.25.2. AQUATIC INVASIVE SPECIES PROGRAMS AND ACTIVITIES

- **Control, Mississippi Department of Wildlife, Fisheries, and Parks.** The Department chemically treats water hyacinth and common salvinia in Mississippi state park waters and fishing lakes and stocks grass carp and salvinia beetles.

- **Monitoring and control, Mississippi Department of Agriculture and Commerce (MDAC), Bureau of Plant Industry, U.S. Department of Agriculture's Animal and Plant Health Inspection Service's (APHIS) Plant Protection and Quarantine (PPQ).** MDAC is assisting APHIS PPQ to monitor and control an infestation of giant salvinia in a private lake. Officials check the lake every 3 months and release salvinia weevils when necessary.

- **Coastal Preserve Program (giant salvinia), Mississippi Department of Marine Resources (MDMR).** Department officials are assessing the possible use of the salvinia weevil to control giant salvinia, which has emerged as an aquatic invasive species (AIS) in the area. Officials are also addressing tallow tree and cogongrass through active surveys for the species and the use of herbicides and mechanical removal for control.

- **Alabama-Mississippi Rapid Assessment Team.** State scientists conduct a 3–5 day survey of all AIS present in the coastal waters of Alabama and Mississippi to establish a baseline for further analysis.

A.25.3. CLIMATE CHANGE CONCERNS

- A recent document from MDMR (Dale A. Diaz and Jeff Clark, *Mississippi Department Of Marine Resources Efforts Related To Aquatic Invasive Species*, Proceedings of the 14th Biennial Coastal Zone Conference, New Orleans, Louisiana, July 2005), states that AIS "[are] a problem because there are many elements in place that make the state susceptible to aquatic invasions," including the following: abundant pathways, such as commercial shipping, heavy recreational watercraft usage, aquaculture, and the ornamental plant trade industry; a subtropical climate with abundant aquatic habitat that is naturally hospitable to AIS; and increased coastal development, which can enhance the establishment of invasive species in areas where habitat has been altered.

A.25.4. CLIMATE CHANGE ACTIONS

- The MDMR document (see CLIMATE CHANGE CONCERNS above) also states, "[the plan will] include sections on the pathways of introduction, education/outreach, prevention, control, eradication, restoration, early detection and rapid response for aquatic invasive species." Mississippi will work with a regional panel to coordinate its activities, and the state will be involved in the Alabama-Mississippi Rapid Assessment Team.

A.25.5. RESEARCH ACTIVITIES AND INFORMATION USED
(None reported.)

A.25.6. RESEARCH NEEDS

- A database of taxonomists who can identify invasive species needs to be developed.[2]

- The use of the salvinia weevil and the potentially negative impacts of its introduction need to be assessed.

- More information on the potential long-term negative effects of control methods is needed.

- More information and expertise on esoteric species is needed.

---

[2]Because the research was completed for this analysis, the need for a database of taxonomists has been fulfilled (see *ANS Task Force Experts Database* at http://www.anstaskforce.gov/experts/search.php).

A.26. SUMMARY OF AQUATIC INVASIVE SPECIES MANAGEMENT IN MISSOURI

A.26.1. AQUATIC INVASIVE SPECIES MANAGEMENT PLAN

Missouri's Aquatic Nuisance Species (ANS) Management Plan was published in August 2005 (see Appendix B, State Aquatic Invasive Species Management Plan Summaries for a general description of the Plan).

A.26.2. AQUATIC INVASIVE SPECIES PROGRAMS AND ACTIVITIES

- **Invasive Species Program, Missouri Department of Conservation (MDC), Fisheries Program.** The program provides public information, and officials are currently enacting the 2005 aquatic invasive species (AIS) Plan. In addition, regulations have been enacted recently that prohibit the use of live bighead and silver carp as bait, create a prohibited species list, and require registration for all sellers of live fish or crayfish as bait. The program is also in the process of developing regulations related to invasive species management (not yet approved).

- **Protect Our Waters Project, MDC, Resource Science Division.** This project, outlined in the Missouri AIS Plan, involves joint work among inter-agency experts on invasive species.

- **Alternate use of redear sunfish for control of snails in aquaculture, MDC, Resource Science Division.** The Division is evaluating the use of redear sunfish, as a substitute for Asian black carp, to control snails in aquaculture ponds.

- **Asian carp heuristic modeling, MDC, Resource Science Division and University of New Orleans.** The project is evaluating a modeling technique to predict the expansion of Asian carp in the Middle Mississippi River system and associated tributary streams.

- **Reeds Canarygrass Management, MDC, Resource Science Division.** The Division is evaluating control of Reeds Canarygrass in wetlands that uses a combination of mechanical and herbicide treatments.

- **Statewide Crayfish Conservation and Management Program, MDC, Resource Science Division.** This program has several components: (1) *Systematic Monitoring Project*—monitors invasions and seeks to set up a long-term monitoring project; (2) *Consulting*—encourages the public to use native species for sale food, or bait, advocates for the addition of the Australian and Rusty Crayfish to the prohibited species list, and works with law enforcement officers to track invasive crayfish, particularly their transport to and from other states; (3) *Stream resource management*—researches inter-species breeding, competition for resources, takeovers of breeding grounds, etc.; (4) *Education*—produces videos, brochures, and articles and gives presentations to school groups; (5) *Working with the bait industry*—built a database of every bait shop in the state (about 400 shops) and found that 90 of these sell crayfish, working on a brochure for the bait and culture industry that shows shops how to identify the five legal crayfish species; (6) *Permits for species collection*—issues permits, usually to teachers, with a requirement that species be released at the same location from which they were obtained.

- **Zebra Mussel Educational Outreach and Monitoring, MDC, Policy Division.** As part of the 100th Meridian Initiative, MDC conducts statewide outreach, including assistance to marine operators who inspect boats for zebra mussels, publication of articles about zebra mussels, supply of information at fairs, outdoor events, and hometown festivals.

- **Invasive Species Management Program, Missouri Department of Natural Resources, Division of State Parks.**

A.26.3.   CLIMATE CHANGE CONCERNS

- Missouri is concerned about the effects of climate change on invasive species in general.

- The state is concerned about increased movement of AIS through interstate commerce and recreation.

A.26.4.   CLIMATE CHANGE ACTIONS

- Missouri has instituted Traveler Information Stations, boat ramp signs, and public-private partnerships address pathways.

A.26.5.   RESEARCH ACTIVITIES AND INFORMATION USED

- Monitoring efforts to track zebra mussels are being implemented.

- Discussions of markets for Asian carp as pet food, oil, consumption, and private use are increasing.

- An ANS workshop on communication strategies for the 2007 North American Fish and Wildlife Conference in Portland, OR was developed.

A.26.6.   RESEARCH NEEDS

- Information on the effects of crayfish on other aquatic species is needed.

- Methods to control crayfish are needed.

- Adequate monitoring and inventories in order to understand the full spectrum of biodiversity in streams are needed.

- Monitoring in order to gauge changes and detect species as soon as they appear needs to be conducted.

A.27. SUMMARY OF AQUATIC INVASIVE SPECIES MANAGEMENT IN MONTANA

A.27.1. AQUATIC INVASIVE SPECIES MANAGEMENT PLAN

Montana's Aquatic Nuisance Species (ANS) Management Plan was published in 2002 (see Appendix B, State Aquatic Invasive Species Management Plan Summaries for a general description of the Plan).

A.27.2. AQUATIC INVASIVE SPECIES PROGRAMS AND ACTIVITIES

- **Montana ANS Program, Montana Department of Fish, Wildlife, and Parks.** The primary goal of the Montana ANS Program is collaboration and coordination with other agencies and other states. The Program consists of five key areas: coordination, education, prevention and control, monitoring, and rapid response. Within the program areas, various activities are being implemented:

  o *Education.* The program conducts education in schools, colleges, and universities, with specialized groups such as Trout Unlimited, and through fishing tournaments, radio stations, and boat launches. The Program also has an education program for professionals. For example, firefighters are targeted because much of Montana's fire equipment is brought in from other states, which subsequently spreads ANS.

  o *Prevention and Control.* The program operates a boat inspection program. This began in 2004, by targeting fishing tournaments on high use waters. In 2005, efforts expanded state-wide, with inspections at more high-use water areas and times. Officials set up angler check stations at major tournaments and water bodies, where anglers must fill out a questionnaire about where they are launching their boats. Cleaning equipment is available to remove debris and sediment, if necessary, before launching. The program also checks for live bait from outside the state and has a hatchery inspection program.

  o *Monitoring.* The monitoring program has inspected all major water bodies for invasive invertebrates and plants. Additionally, a whirling disease distribution study has been underway for several years. Officials also test fish for diseases and map their spread. There is also an ongoing distribution study of New Zealand Mud Snails, with plans to test all fishing access sites. The resulting information is entered into a national Internet database.

  o *Rapid Response.* Montana officials have a rapid response plan in place for zebra mussels upon detection, with different plans based on general scenarios. They are also mapping national statistics to identify and monitor the most likely areas where mussels might be introduced.

A.27.3. CLIMATE CHANGE CONCERNS

(None reported.)

A.27.4. CLIMATE CHANGE ACTIONS

(None reported.)

A.27.5. RESEARCH ACTIVITIES AND INFORMATION USED

- Montana's aquatic invasive species (AIS) program researched the effectiveness of ANS Program outreach tools.

- The state's AIS program completed a specific study in 2006 to examine the effectiveness of Traveler Information Systems on public outreach.

- Montana's AIS program is conducting surveys annually to identify transport patterns within the state in order to help identify bodies of water at highest risk of introduction.

A.27.6. RESEARCH NEEDS

- Montana needs to undertake risk assessments for the establishment of other aquatic invasives within the state. The state should determine which bodies of water are at highest risk of establishment and which species are most likely to become established.

A.28. SUMMARY OF AQUATIC INVASIVE SPECIES MANAGEMENT IN NEBRASKA

A.28.1. AQUATIC INVASIVE SPECIES MANAGEMENT PLAN
No plan available.

A.28.2. AQUATIC INVASIVE SPECIES PROGRAMS AND ACTIVITIES

- **Coordination, Nebraska Game and Parks Commission (NGPC), Fisheries Division.** The Division conducts education and outreach, as well as some control for the common carp.

- **Noxious Weed Program, Nebraska Department of Agriculture (NDA), Bureau of Plant Industry.** This Program conducts several activities: oversight of weed control superintendents around the state; training and education of personnel on the Nebraska Noxious Control Act, infestations, and control methods; dissemination of information and educational campaigns; designation of noxious weeds and their control measures; collection of information from counties regarding presence of noxious weeds; and cooperation with federal and state agencies.

- **Douglas County Noxious Weed Control Authority, Douglas County Environmental Services.**

- **Lancaster County Weed Control Program, Lancaster County Weed Control Authority.**

- **Lower Platte Weed Management Area, a partnership among the county weed control boards (Butler, Cass, Colfax, Dodge, Douglas, Lancaster, Platte, Sarpy, Saunders, Seward, and Washington), NGPC, NDA.** This program surveys, and monitors for and controls purple loosestrife in the Platte River Drainage. Officials have surveyed nearly 100 miles of the Platte River and treated nearly 75% of the infestations by chemical or insects releases. Continued monitoring and control is planned.

- **Twin Valley Weed Management Area (TVWMA), a partnership among the county weed control boards (Adams, Clay, Fillmore, Franklin, Furnas, Harlan, Kearney, Nuckolls, Thayer, and Webster counties), NDA, University of Nebraska Extension, and Board of Educational Lands and Funds.** TVWMA facilitates coordination among land managers and landowners to identify and manage noxious and invasive plant problems and conducts outreach and education.

- **Nebraska Weed Control Association.** This is a forum where superintendents can exchange information about noxious weeds.

- **Adopt-A-Stream program, Nebraska Wildlife Federation.** This program teaches local volunteers how to conduct chemical and biological monitoring.

A.28.3. CLIMATE CHANGE CONCERNS

- Increased drought caused by changes in climate may cause purple loosestrife and Phragmites populations to increase drastically.

- Warmer temperatures may affect some species, but not others.

A.28.4. CLIMATE CHANGE ACTIONS

(None reported.)

A.28.5. RESEARCH ACTIVITIES AND INFORMATION USED

- Nebraska is developing a strategy to eliminate or reduce purple loosestrife through the use of mechanical (digging), chemical (herbicides), or biological (insect) controls.

- Chemical experiments are being conducted on Phragmites by spraying habitat via helicopter over 80 acres along the river. Grazing cows and goats are also being used as a trial method to control Phragmites.

A.28.6. RESEARCH NEEDS

- Information about Asian carp is needed.

- Identification of native and non-native species is needed.

- New techniques for more effective or selective control and herbicides need to be developed.

- An understanding of where purple loosestrife seeds will be disseminated and where it may reappear is needed.

- More knowledge about the anatomy and botany of invasive plants is needed.

A.29. SUMMARY OF AQUATIC INVASIVE SPECIES MANAGEMENT IN NEVADA

A.29.1. AQUATIC INVASIVE SPECIES MANAGEMENT PLAN

No plan available.

A.29.2. AQUATIC INVASIVE SPECIES PROGRAMS AND ACTIVITIES

- **Lake Tahoe Basin Weed Coordinating Group, Nevada Department of Agriculture and University of Nevada Cooperative Extension.** The group conducts and/or encourages the following activities: (1) *Aquatic weed work at Lake Tahoe*—the initial phase includes public education and outreach (posting signs and distributing information to boaters, asking for boat cleaning, and disseminating flyers at Forest Service leaseholders' homes to alert of potential spread); (2) *Voluntary removal of pondweed* by landowners and managers (curlyleaf pondweed is on the Tahoe Priority Weeds List); (3) *Eurasian milfoil control* at the south end of the lake through mechanical weed removal and induced water temperature changes to prevent spread and growth of weed; and (4) *Pilot project for weed removal*—California Department of State Lands has spearheaded a pilot project to use diver-assisted weed removal at the south end of the lake. A Bureau of Reclamation grant will allow the work to continue and expand through 2010. The use of bottom barriers is also being investigated.

- **Control of tall whitetop and tamarisk, Nevada Department of Conservation and Natural Resources, Muddy River Regional Environmental Impact Alleviation Committee, Southern Nevada Water Authority, and others.** Nevada uses inmate labor crews to control tall whitetop and tamarisk. The strategy consists of mechanical control (cutting down plants with chainsaws) and herbicide application to stumps. Many of the Nevada Cooperative Weed Management Areas (CWMAs) also participate in tamarisk and other aquatic invasive species removal.

- **Biocontrol of tamarisk, University of Nevada-Reno, in cooperation with U.S. Department of Agriculture.** The "Saltcedar Biological Control Consortium," a multi-agency and multi-partner effort that includes private interests such as the Cattlemen's Association and conservation groups such as The Nature Conservancy, conducts tamarisk research, including a biocontrol project using weevils.

- **Chemical removal of undesirable species, Nevada Department of Wildlife (NDOW).** The Department removes some non-native game fish to perpetuate native species (cutthroat trout, bull trout, and other trout species) that are currently or potentially threatened or endangered. NDOW treats the water body for two consecutive years with piscicides and then restocks with native fish.

- **Invasive Species Management Plans, NDOW, Wildlife Management Areas.** Each Wildlife Management Area, established to protect habitats and biodiversity, is developing an Invasive Species Management Plan. The department continues to work closely with CWMAs to control invasive weeds on state-owned Wildlife Management Areas.

Grazing, herbicidal spraying, and biocontrols have been implemented in many of the areas.

- **Database of Invasive Plant Mapping Data, Nevada Natural Heritage Program.** Nevada Natural Heritage Program, in cooperation with the Nevada Department of Agriculture, is creating a database management and mapping position to keep track of all of Nevada's Invasive Plant Mapping Data, including aquatic plants. The data will be gathered by CWMAs and various agencies and organizations.

A.29.3. CLIMATE CHANGE CONCERNS

- Although weevils have effectively defoliated the tamarisk trees numerous times, they do not do well in southern Nevada. This may be linked to temperature, since weevils may be heat-sensitive. Researchers may try to obtain more heat-tolerant weevils from their source location.

- Climate change could increase demand on water resources, and because invasive species (such as tamarisk) deplete the water supply, invasive species could become a greater problem as a result of climate change.

A.29.4. CLIMATE CHANGE ACTIONS
(None reported.)

A.29.5. RESEARCH ACTIVITIES AND INFORMATION USED
(None reported.)

A.29.6. RESEARCH NEEDS

- Cooperative Extension education and outreach need to be developed.

- A tamarisk-eating weevil that is effective in the south need to be identified.

- More effective herbicides for treating tall whitetop and tamarisk need to be developed.

- More information about the effect of chemicals on non-target species such as macroinvertebrates and their recovery needs to be gathered.

- More research on other biocontrols for invasive species needs to be conducted.

A.30. SUMMARY OF AQUATIC INVASIVE SPECIES MANAGEMENT IN NEW HAMPSHIRE

A.30.1. AQUATIC INVASIVE SPECIES MANAGEMENT PLAN

No plan available.

A.30.2. AQUATIC INVASIVE SPECIES PROGRAMS AND ACTIVITIES

- **New Hampshire Exotic Aquatic Plant Program, New Hampshire Department of Environmental Services (NHDES).** The primary purpose of the program is to "prevent the introduction and further dispersal of exotic aquatic weeds and to manage or eradicate exotic aquatic weed infestations in the surface waters of the state." The program focuses on submerged exotic aquatic plants, including variable milfoil (*Myriophyllum heterophyllum*), Eurasian watermilfoil (*Myriophyllum spicatum*), fanwort *(Cabomba caroliniana)*, Brazilian elodea (*Egeria densa*), Hydrilla (*Hydrilla verticillata*), and water chestnut (*Trapa natans*), among other species. The program has five focus areas: (1) *Prevention* of new infestations; (2) *Monitoring* for early detection of new infestations to facilitate rapid control activities; (3) *Control* of new and established infestations; (4) *Research* towards new control methods with the goal of reducing or eliminating infested areas; and (5) *Regional cooperation.* The program is funded through the collection of a $5 fee derived from New Hampshire boat registrations. For each $5 collected, $4.50 is dedicated to tasks and projects associated with exotic aquatic plants. The program's establishing statutes also list 27 prohibited aquatic plants and associated species.

- **Lake Host Program, NHDES and New Hampshire Lakes Association.** The New Hampshire Lakes Association, under a grant appropriation from NHDES, hires summer staff to inspect aquatic recreational gear, such as boats, trailers, and personal water craft, for aquatic weeds at public access water sites across the state. Staff also distribute information and maps on exotic aquatic plant infestations. If detected, aquatic invasive species are sent to NHDES, which posts an online notice. The Lakes Association also educates boaters about self-inspections.

- **Weed Watcher Program, NHDES.** NHDES trains volunteers to monitor water bodies for any new growths of exotic aquatic plants. If volunteers find aquatic weeds, they cull them and send to NHDES. New infestations are assessed and removed using hand-culling, bottom barriers, or herbicides following the principles of Integrated Pest Management. As of 2006, NHDES monitoring activities included surveys of over 300 lakes, with over 600 trained Weed Watchers actively monitoring water bodies across the state.

- **Milfoil Control, Squam Lakes Association (SLA) and NHDES.** SLA organizes trained volunteers to conduct surveys, remove fragments, and pull rooted variable milfoil. Questionable specimens are sent to SLA for identification and NHDES is notified of new infestation sites. Control is possible, but eradication is not. Since the discovery of milfoil, SLA has been working with the NHDES to develop management alternatives for the infestations. An ad hoc milfoil task force has been formed between SLA, the marina operators, and SLA's consulting ecologist. In 2006, NHDES granted research funds to Plymouth State University, which was working in partnership with SLA to conduct research projects in portions of Squam Lake. Research examined the impacts of a 2,4-D herbicide treatment on the benthic fauna of the lake (including macroinvertebrates).

- **Exotic Species Management, NHDES.** NHDES annually coordinates the management of exotic aquatic plants in 15–20 water bodies. Variable milfoil, and more recently, fanwort, are the two plants that are most often the target of these control practices. Control measures for new, small infestations include hand pulling or benthic barriers, and may include designation of a Restricted Use Area in the vicinity of the infestation. Larger, established infestations are usually controlled with herbicides.

- **Milfoil Research (general and specific), NHDES.** (See RESEARCH ACTIVITIES AND INFORMATION USED below.)

A.30.3. CLIMATE CHANGE CONCERNS REPORTED BY STATE PERSONNEL

- Many species that NHDES encounters are southern species from South America and Africa that have migrated northward and are surviving in this cooler climate. Plants may be adapting, or climate change may be lengthening the growing seasons. With recent mild winters, plants may have the opportunity to gain a foothold. Plants of concern because of climate change include: giant salvinia, water hyacinth, and water lettuce. These are warm water southern species that can currently survive the summer, but not the winter.

A.30.4. CLIMATE CHANGE ACTIONS

- NHDES has just expanded the list of prohibited species to include a total of 27 plants. This was done to account for the northward migration of southern species. NHDES hopes that by listing plants as prohibited, they will not be circulated in the state through the aquatic plant industry, thereby lessening their likelihood of introduction through that avenue. Neighboring states to New Hampshire are also following suit.

A.30.5. RESEARCH ACTIVITIES AND INFORMATION USED

- Specific strategies are being developed for aquatic herbicide use that incorporate plant phenology, water quality, and treatment timing for optimal, cost-effective, and selective control of variable milfoil.

- The plant and nematode communities, along the with water chemistry and sediment conditions, associated with variable milfoil in its native range and in New Hampshire lakes are being compared and characterized, and possible plant-nematode association for biocontrol of variable milfoil are being sought.

- Effects of chemical and physical properties on variable milfoil are being evaluated, an effective monitoring tool is being developed, and optimal aquatic habitat for milfoil establishment and growth is being determined. Geophysical and vegetation surveys and water quality sampling also are being conducted, and data will be integrated.

- Lake attributes that influence distribution of native and non-native milfoils are being identified, and multivariate statistics and logistic regression are being used to determine whether invasive milfoil species are correlated with chemical, morphological, biological, and spatial characteristics of New Hampshire lakes. Results of this study will identify classes of lakes that may be susceptible to invasion.

- The effects of water and sediment chemistry, sediment physical properties, number and size of contiguous wetlands, and watershed geology on variable milfoil abundance or presence/absence are being investigated.

- The Plant Replacement Program works to establish a native, non-nuisance assemblage dominated by low-growing species. This effort involves both removal of the current dominant milfoil population over a target area early in the growing season and planting or seeding with the desired species.

- NHDES is studying the effectiveness of the herbicide 2-4D. NHDES did intensive GIS mapping of a lake and arranged 2-4D pellets in a consistent manner to target plants exactly where they are growing and to ensure that the chemical goes directly to the plants. NHDES is monitoring to ensure effectiveness.

- NHDES partnered with Plymouth State University to conduct a research project on the effects of a 2-4D treatment on the chemistry, biology, and ecology of a small portion of Squam Lake. Data from pre- and post-herbicide treatment are included in the study. Data from this study were to be released in the fall of 2007.

A.30.6. RESEARCH NEEDS

- Variable milfoil research is needed.

- Chemical and biocontrol methods are needed.

- The biology and ecology of plants and what makes them invasive, as well as the habitat characteristics that invasive plants favor need to be researched.

A.31. SUMMARY OF AQUATIC INVASIVE SPECIES MANAGEMENT IN NEW JERSEY

A.31.1. AQUATIC INVASIVE SPECIES MANAGEMENT PLAN
No plan available.

A.31.2. AQUATIC INVASIVE SPECIES PROGRAMS AND ACTIVITIES

- **Liberty State Park Project: Interior Restoration, New Jersey Department of Environmental Protection (NJDEP).** This project derives from a legal settlement of a chromium case and restoration involves Phragmites control by tidal flushing. NJDEP eradicates the species by removing the soil from an area around the plant, uprooting the plant, and filling the holes with "clean" sand.

- **Lower Cape May Meadows Environmental Restoration Project, NJDEP.** NJDEP controls Phragmites using fill and herbicides and controls purple loosestrife using beetles and tidal flushing.

- **Partners for Fish and Wildlife Program: purple loosestrife control, U.S. Fish and Wildlife Service, New Jersey Department of Agriculture, NJDEP, Division of Fish and Wildlife, Endangered and Nongame Species Program.** The Program controls purple loosestrife on private lands.

- **New Jersey Invasive Species Council.** The Council was created under a 2004 executive order to create a state invasive species management plan and to undertake a set of tasks to control and eradicate invasive species in the state. Representatives on the Council come from the NJDEP, Department of Agriculture, Department of Transportation, Commerce and Economic Growth Commission, conservation organizations, agricultural sector, nursery and landscape sectors, New Jersey Agricultural Invasive Species Council, academia, and the general public.

- **Wetlands enhancement in the New Jersey Meadowlands, New Jersey Meadowlands Commission.** Wetland enhancement projects for three sites in the New Jersey Meadowlands area include control and management of numerous invasive species.

A.31.3. CLIMATE CHANGE CONCERNS REPORTED BY STATE PERSONNEL
(None reported.)

A.31.4. CLIMATE CHANGE ACTIONS
(None reported.)

A.31.5. RESEARCH ACTIVITIES AND INFORMATION USED

- Standardized monitoring protocols for restoration projects are being established.

A.31.6. RESEARCH NEEDS

(None reported.)

A.32. SUMMARY OF AQUATIC INVASIVE SPECIES MANAGEMENT IN NEW MEXICO

A.32.1. AQUATIC INVASIVE SPECIES MANAGEMENT PLAN

New Mexico's Aquatic Nuisance Species Management Plan is currently under review by the New Mexico Department of Game and Fish (NMDGF).

A.32.2. AQUATIC INVASIVE SPECIES PROGRAMS AND ACTIVITIES

- **Lower Rio Grande Salt Cedar Control Project, New Mexico Association of Conservation Districts.** The project includes the following: eradication efforts; development of management and native vegetation restoration plans; hearings to receive public input on the plans; aerial spraying by helicopter or ground application with prior public notice; and monitoring and evaluation of the effects of control on wildlife, water quality, vegetation, and soil health.

- **Salt Cedar Task Force, New Mexico Environment Department.**

- **Strategy for Long-Term Management of Exotic Trees in Riparian Areas for New Mexico's Five River Systems, New Mexico Interagency Weed Action Group.** Efforts include prevention, early detection and mapping, timely control, and adaptive management. Control includes manual removal, selective mechanical grubbing, low-volume basal bark herbicide application, cut-stump herbicide application, foliar herbicide application, and aerial herbicide applications for Russian olive, salt cedar, and Siberian elm. This strategy encompasses ecosystem impacts, including stream bank stabilization, increased evapotranspiration, altered fire regimes, salt uptake, and decreased native biodiversity.

- **Native Trout Management, NMDGF, U.S. Department of Agriculture Forest Service, National Park Service, U.S. Fish and Wildlife Service (USFWS), New Mexico Interstate Stream Commission, and private groups.** Together, these organizations seek to halt and/or reverse the invasion of non-native trout and its effects on native cutthroat trout. Most work involves managing non-native trout populations through electrofishing or by physical removal. NMDGF also installs migration barriers to prevent the invasion of currently un-invaded streams. They also conduct chemical treatment and restoration of the gila trout, which is protected under both state and federal law.

- **Whirling Disease Program, NMDGF, USFWS.** The NMDGF has implemented a statewide monitoring program to track the status of whirling disease in infested and negative salmonid populations using GIS-based mapping. The program also tests for presence of whirling disease in hatchery stock and in native and managed trout populations.

- **Golden Algae Monitoring Program, NMDGF.** The NMDGF is conducting statewide monitoring of the golden algae to determine the effects of algal blooms on zooplankton, fish communities, and aquatic macroinvertebrates. The measurement of physicochemical parameters will serve to develop predictors for blooms and toxic events and to prescribe management actions to maintain sport fisheries, native fish communities, and aquatic macroinvertebrates.

- **San Juan River Non-native Fish Removal Program, NMDGF, USFWS, Bureau of Reclamation, Utah Division of Wildlife Resources**. The Program, a collaborative efforts since 2001, restores the native fish of the San Juan River, including physical removal of non-native piscivores and common carp.

- **Non-native Crayfish Survey, NMDGF.** Since 1991, the NMDGF has been actively documenting the statewide occurrence of non-native crayfish.

- **Zebra Mussel Monitoring, USFWS, U.S. Army Corps of Engineers, New Mexico State Parks.** Zebra mussel monitoring was initiated in 2005 at three state parks (Conchas Lake, Heron Lake, Elephant Butte) and two sites on the Rio Chama.

- **Chytrid Fungus Monitoring, NMDGF, Western New Mexico State University, Pisces Molecular (Boulder, CO).** Chytrid fungus infections, implicated in the decline of amphibians worldwide, are known to occur in four species of anurans and one salamander in New Mexico. Using molecular genetic techniques, collaborative efforts are ongoing to survey New Mexico for incidence of occurrence in other amphibian taxa.

A.32.3. CLIMATE CHANGE CONCERNS

- Climate change could have significant effects on native fish. An increase of even a few degrees in water temperature would lead to loss of habitat and species. Non-native trout with higher tolerance to warmer water temperatures and degraded water quality would be at an advantage.

A.32.4. CLIMATE CHANGE ACTIONS
(None reported.)

A.32.5. RESEARCH ACTIVITIES AND INFORMATION USED

- Statewide surveys for non-native crayfish are being conducted.

A.32.6. RESEARCH NEEDS

- The upper temperature tolerance of fish, the impact of varying degrees of water quality on fish, the mechanisms through which non-native trout out-compete or displace native trout, and the native trout's life history characteristics need to be researched.

- The effects of piscicides on amphibians and mollusks, particularly the early life stages of tadpoles and aquatic insects, need to be studied in more detail.

- A method for field detection of antimycin in streams needs to be developed.

- Research on antimycin's persistence time in waters of different qualities is needed.

- Statewide surveys for non-native crayfish should be continued in order to develop a database and to synthesize results for directing management strategies.

- Research on the effects of non-native crayfish on aquatic ecosystems need to be conducted.

- Influences of atmospheric conditions on golden algae blooms need to be investigated.

- Statewide surveys of amphibians for chytrid fungus need to be expanded.

A.33. SUMMARY OF AQUATIC INVASIVE SPECIES MANAGEMENT IN NEW YORK

A.33.1. AQUATIC INVASIVE SPECIES MANAGEMENT PLAN

New York's Nonindigenous Aquatic Species Comprehensive Management Plan was published in 1993 (see Appendix B, State Aquatic Invasive Species Management Plan Summaries for a general description of the Plan).

A.33.2. AQUATIC INVASIVE SPECIES PROGRAMS AND ACTIVITIES

- **New York State Invasive Species Task Force.** The Task Force is composed of multiple state agencies and nongovernmental organizations and is jointly chaired by the New York Department of Environmental Conservation and the New York Department of Agriculture and Markets. The original function of the Task Force was to evaluate the spectrum of invasive species issues and to make recommendations to the legislature and Governor as to how the state should address the issue. The Task Force has completed this report and now works to implement the proposed recommendations.

- **Purple Loosestrife Biocontrol Program, Cornell University, Ecology and Management of Invasive Plants Program.** The Program releases biocontrol insects at over 4,000 sites across the country. It is trying to determine why treatment has succeeded in some areas and not others.

- **Phragmites, water chestnut, Japanese knotweed biocontrol research, Cornell University, Ecology and Management of Invasive Plants Program.** Scientists are researching biocontrol options.

- **Lake Services Section, New York Department of Environmental Conservation (NYSDEC), Division of Water.** The Division provides local assistance grants for aquatic plant control. It operates a volunteer program to teach plant identification and how to collect and submit samples, conducts plant research and surveys in Lake George, and engages in public outreach through conferences, lake association meetings, site visits, and management activities.

- **NYSDEC, Division of Fish, Wildlife, and Marine Resources.** The Division undertakes local management of aquatic invasive species (AIS) and is modifying regulations to prevent introduction of Chinese mitten crab. It also has a program for hand-harvesting water chestnuts and monitoring AIS, such as round goby, spiny water fleas, and zebra mussels, and their effects (which includes the ecological effects of zebra mussels in eight Finger Lakes). Finally, the Division administers a $1 million grant program for AIS eradication projects. (In FY2005, 32 grants were funded. The program will continue in FY 2006, but the funding will be shared with a terrestrial invasive species eradication grant program that is currently under development.)

- **Sea Lamprey Control, NYSDEC, Bureau of Fisheries.** The bureau undertakes sea lamprey control using chemicals and migration barriers.

- **Monitoring Program: zebra mussels, quagga mussels, and round goby NYSDEC, Bureau of Fisheries.** The bureau monitors for Type E botulism and collects dead bodies. Control is not feasible.

- **NYSDEC, Division of Fish, Wildlife, and Marine Resources, Region 5 and Region 6.** The Division protects ponds that are habitat for unique strains of native Adirondack brook trout from species such as yellow perch.

- **Aquatic Plant Harvesting, Finger Lakes-Lake Ontario Watershed Protection Alliance (FL-LOWPA).** The Alliance conducts mechanical harvesting in multiple counties at multiple sites for aquatic plants including Eurasian watermilfoil (EWM), muskgrass, and water chestnut. Some county programs have volunteer training and opportunities.

- **Invasive Species Initiative, FL-LOWPA, Hamilton County Soil and Water Conservation District (SWCD).** The District distributes educational materials, including fact sheets, brochures, and signs, and is developing and encouraging volunteer monitoring for invasive aquatic plants, which will provide assistance to several lake associations.

- **Evaluating Alternative Control Strategies for Invasive Aquatic Plants, FL-LOWPA, Madison County Planning Department, in conjunction with SUNY Oneonta and Cornell University.** With the goal of formulating a control strategy, the group is examining the impact of fish communities on EWM herbivores.

- **Zebra Mussel Monitoring in Eaton Brook Reservoir and Downstream Tributaries, FL-LOWPA, Madison County Planning Department.** These entities are monitoring the zebra mussel population established in the reservoir, because it is a tributary to the Susquehanna River which empties into the Chesapeake Bay where the zebra mussels are not yet established.

- **Monitoring and Research, FL-LOWPA, Steuben County SWCD, in cooperation with Cornell University Experimental Ponds Program.** The District is conducting research on the presence and impact of the European aquatic moth (an exotic species that feeds on EWM).

- **The Milfoil Project (Weevil Control Program), Lake Bonaparte Conservation Club.** The Club is conducting milfoil control using weevils.

- **Milfoil Control, Upper Saranac Lake Foundation.** The town contracted with divers to hand-cull milfoil in the Upper Saranac Lake.

- **Research, Cornell University, Research Ponds Facility.** Researchers are monitoring and managing aquatic plant communities throughout the northeast and New York State and demonstrating physical, biological, and chemical control methods for aquatic nuisance species.

- **Research, Cornell University, Aquatic Research Facility.** Researchers are contributing to a 50+ year, long-term dataset on Oneida Lake, New York that includes information on invasives and an aquatic foodweb ranging from nutrients to top predators. An experimental facility examines foodweb impacts of New York invasives in research settings ranging from small-scale aquaria to large-scale mesocosms.

- **Water Chestnut Control, State University of New York Oneonta Biological Field Station in cooperation with state agencies, nongovernmental organizations, and private stakeholders.** Focus is on nutrient export associated with control activities.

- **Japanese Knotweed Initiative, Delaware River Invasive Plant Partnership (DRIPP).** DRIPP develops educational brochures and works with local community volunteer sites to provide best scientific guidelines and demonstration control sites (showcasing repeated cutting to keep knotweed under control and prevent it from spreading).

- **Japanese Knotweed Study, New York City Department of Environmental Protection, in conjunction with Green County SWCD.**

- **Delaware River Invasive Plant Partnership, States of Delaware, New Jersey, New York, and Pennsylvania.**

A.33.3. CLIMATE CHANGE CONCERNS

- With climate change, purple loosestrife could move further north, where biocontrol insects may not survive. The range of plants and insects may shift and southern invasive species could move into New York.

- Water hyacinth is sold all over the state. Currently, it does not survive the winter in New York. However, this could change with climate change.

- Climate change could cause changes in the native vegetation and, depending on the rate at which that happens, could lead to more pest problems.

A.33.4. CLIMATE CHANGE ACTIONS
(None reported.)

A.33.5. RESEARCH ACTIVITIES AND INFORMATION USED

- Dynamics of decomposition for invasive weeds (Phragmites) and native cattails (*Typha*) are being compared in order to determine the benefit of restoration efforts.

- The way in which nutrient level changes and exotic mussels affect the Lake Erie food web and the fish community are being examined.

- A genetic probing technique is being developed that will quickly screen water samples for zebra mussel veligers.

- The role of embayments and inshore areas as nursery grounds for alewife and other species is being examined.

- The influence of zebra mussels in metal cycling in freshwater ecosystems is being examined and whether zebra mussels may serve as bioindicators for the presence of toxic metals in freshwater systems is being investigated.

- The effects of zebra mussels on the spawning shoals of walleye and lake trout are being studied.

- Plots of Japanese knotweed are being treated and plotted in order to test three control methods: (1) repeated cutting; (2) herbicide injections; and (3) limited excavation with replanting.

- A non-herbicide approach for treatment of knotweed is being identified.

A.33.6. RESEARCH NEEDS

- Plants not currently targeted for biocontrol, such as curlyleaf pondweed need to be researched.

- More information about how to restore wetlands after biocontrol need to be attained.

- Studies should be conducted to determine whether biocontrol organisms identified overseas are specific enough for the species that are being targeted in New York (Knotweed, Water Chestnut, and Phragmites) and whether these biocontrols can be introduced safely into North America.

- Economic and agricultural impacts of invasive species need to be demonstrated.

- Information on biocontrol (predators, pests, diseases) for sea lampreys and on how knotweed affects aquatic species needs to be attained.

- Research is need on mussel control methods, especially for the quagga mussel.

A.34. SUMMARY OF AQUATIC INVASIVE SPECIES MANAGEMENT IN NORTH CAROLINA

A.34.1. AQUATIC INVASIVE SPECIES MANAGEMENT PLAN

No plan available.

A.34.2. AQUATIC INVASIVE SPECIES PROGRAMS AND ACTIVITIES

- **Aquatic Weed Control Program, North Carolina Department of Environment and Natural Resources (NC DENR), Division of Water Resources.** The Division removes invasive species through different on-the-ground methods: (1) *Physical control*—water level manipulation, deepening near-shore areas; (2) *Mechanical control*—removal of weeds with hand tools; (3) *Biocontrol*—herbivorous fish or insects that attack specific weeds; and (4) *Chemical control*—herbicides approved by the U.S. Environmental Protection Agency for aquatic use. The Division assists local governments by: *Providing* cost-share grants for qualifying projects (municipalities, counties, soil and water conservation districts, government agencies, and public utilities are eligible for assistance); *Assessing* sites and providing recommendations when control efforts are needed; and *Identifying* aquatic weed infestations. The Division also assists the general public by providing free evaluations of aquatic weed problems in private waters and conducting public outreach and education on invasive aquatic weeds. The Division's species-specific work includes Salvinia—experimenting with the host-specific Brazilian weevil (*Cyrtobagous salviniae*) to control giant salvinia and herbicides; Hydrilla—control using sterile grass carp (only sterile "triploid" grass carp may be legally introduced into state waters); herbicides; water draw-downs; and mechanical removal; Alligatorweed—control using herbicides and flea beetles; Parrotfeather—control using triploid grass carp; Creeping water primrose—control using herbicides; Eurasian watermilfoil—biocontrol and herbicides. (Note—parrotfeather, water lettuce, and water hyacinth were added to NC DENR list of noxious aquatic weeds in 2006.

- **Weed Regulatory Services, North Carolina Department of Agriculture and Consumer Services, Plant Industry Division, Plant Protection Section.** Giant salvinia-related work includes active surveys; physical removal; and experimentation with biocontrol (releasing salvinia weevils) in cooperation with the Giant Salvinia Task Force (GSTF). *Lythrum salicaria* (semi-aquatic) work includes surveys and physical and chemical removal. The Commissioner of Agriculture may also regulate the importation, sale, use, culture, collection, transportation, and distribution of a noxious aquatic weed as a plant pest under the General Statutes of North Carolina (see Chapter 106, Article 36).

- **GSTF, a cooperative effort by state, local, and federal agencies and private landowners.** The GSTF conducts the following activities: (1) Uses chemical and biocontrols in areas where giant salvinia has established (herbicides account for 95% of control efforts); (2) Surveys areas adjacent to infestation for evidence of giant salvinia establishment; (3) Responds to reports from around the state of giant salvinia establishment. Within 24 hours of a call, the Task Force assesses the site and arranges for control treatments if salvinia is found.

- **North Carolina State University Aquatic Weed Management Program.** This program conducts research and outreach activities related to invasive plant management on aquatic and non-cropland sites. Activities include the following: *Evaluation* of chemical, biological, physical, mechanical, and other methods of controlling invasive plants; *Determination* of biological and ecological characteristics of invasive plants that contribute to spread, establishment, and management; *Dissemination* of current information to managers, government employees, and others related to management of invasive plants; and *Interaction* with government agencies and private entities to improve management of invasive plants.

A.34.3. CLIMATE CHANGE CONCERNS REPORTED BY STATE PERSONNEL

- Biocontrol is being used for alligatorweed works better in warmer winters.

- Water hyacinth is a problem only in the southeast corner of the state, but this is also the warmest region.

- Air and water temperature monitoring at some sites shows that giant salvinia is surviving at much colder temperatures than the literature reports.

A.34.4. CLIMATE CHANGE ACTIONS
(None reported.)

A.34.5. RESEARCH ACTIVITIES AND INFORMATION USED
(None reported.)

A.34.6. RESEARCH NEEDS

- More information on the best way to control hydrilla is needed (herbicides vs. grass carp).

- More information needs to be collected on the biology and ecology of invasive species (i.e., seed longevity) that would help improve control methods.

A.35. SUMMARY OF AQUATIC INVASIVE SPECIES MANAGEMENT IN NORTH DAKOTA

A.35.1. AQUATIC INVASIVE SPECIES MANAGEMENT PLAN

North Dakota's Aquatic Nuisance Species (ANS) Management Plan was published in 2005 (see Appendix B, State Aquatic Invasive Species Management Plan Summaries for a general description of the Plan).

A.35.2. AQUATIC INVASIVE SPECIES PROGRAMS AND ACTIVITIES

- **Lake Oahe Salt Cedar Task Force and Lake Sakakawea Salt cedar Task Force.** These Task Forces are federal, state, and local partnerships that conduct surveys along Yellowstone River and Lake Sakakawea. Thousands of acres have been surveyed and hundreds of acres have been treated. Early detection and rapid response is the policy of all agencies and organizations for combating salt cedar in the state. Because of this, infested acres have remained low due to the herbicide treatments.

- **Western North Dakota Weed Management Group (encompasses the Little Missouri River from the South Dakota border to Lake Sakakawea, the Lake Sakakawea Saltcedar Task Force, and the recently formed Lake Oahe Saltcedar Task Force).**

- **Purple Loosestrife Weed Management Groups, county/state/federal agencies and private individuals and organizations.** The Lower Sheyenne Purple Loosestrife Project has surveyed and treated the species in the Sheyenne River, from the Bald Hill Dam to the Red River through Fargo. The project has also conducted plant exchanges (garden purple loosestrife for Liatrus), as well as developing, printing, and distributing informative table place mats, table tents, and invasive ornamentals brochures. These items have been shared and distributed statewide in an effort to control and prevent the spread of purple loosestrife and other ornamental invasives. The Souris River Purple Loosestrife Weed Management Group has surveyed and treated the species from Minot, ND to the Canadian Border. They have also had exchange programs. Both working groups have also utilized biocontrol insects and actively surveyed for salt cedar while surveying and treating purple loosestrife.

A.35.3. CLIMATE CHANGE CONCERNS
(None reported.)

A.35.4. CLIMATE CHANGE ACTIONS
(None reported.)

A.35.5. RESEARCH ACTIVITIES AND INFORMATION USED

(None reported.)

A.35.6. RESEARCH NEEDS

- The length of seed viability of salt cedar at northern latitudes and climates needs to be researched. This information would be invaluable in making salt cedar management plans. Field observations by weed managers show that seed is viable much longer in our colder climates than where prior seed viability research was conducted.

- Research on the mechanism of spread of salt cedar is needed. Anecdotal evidence points towards waterfowl and wind as being primary means of salt cedar spread. This research data would assist weed managers in concentrating their survey efforts and resources in those areas most likely to be infested.

- Research should be conducted on ability of ANS to be transported to North Dakota and the likelihood that they will become established in state waters. The study should include a risk assessment based on pathways information, frequency of movement into the state, and suitable habitat availability.

A.36. SUMMARY OF AQUATIC INVASIVE SPECIES MANAGEMENT IN OHIO

A.36.1. AQUATIC INVASIVE SPECIES MANAGEMENT PLAN

Ohio's has a Comprehensive Management Plan for aquatic invasive species (AIS) management (publication date not available) (see Appendix B, State Aquatic Invasive Species Management Plan Summaries for a general description of the Plan).

A.36.2. AQUATIC INVASIVE SPECIES PROGRAMS AND ACTIVITIES

- **Aquatic Nuisance Species Program, Ohio Department of Natural Resources (ODNR), Division of Wildlife.** The Division conducts control efforts for Phragmites, purple loosestrife (including biocontrol), reed canary grass, and flowering rush on its wildlife areas statewide. The Division is also developing a comprehensive AIS Web page and operates a monitoring program to survey for new AIS introductions as well as existing populations.

- **Invasive Species Control and Management, ODNR, Division of Parks and Recreation.** The Division engages in control of invasive species, including Phragmites, purple loosestrife, and milfoil. Control methods vary based on area, need, and funding and include herbicides for Phragmites and loosestrife (spraying Rodeo™ and mowing); disking certain dry areas to destroy roots and reseed with native marsh grasses; and water drawdown to flood out Phragmites (however, this can lead to invasions from exposed dirt).

- **Invasive species control in state preserves, ODNR, Division of Natural Areas and Preserves.** The Division has management plans for each site. Each plan has a policy statement regarding treatment of problematic non-native flora. Guidelines call for manual removal, burning, and herbicide treatment. Plans also include provisions for monitoring and assessment to determine the extent of growth and nature of the disturbance. Plans are tailored to the specific preserve or area and prescribe the treatment appropriate for each species depending upon the habitat type, extent of invasion, and management goals for the area.

A.36.3. CLIMATE CHANGE CONCERNS
(None reported.)

A.36.4. CLIMATE CHANGE ACTIONS
(None reported.)

A.36.5. RESEARCH ACTIVITIES AND INFORMATION USED
(None reported.)

A.36.6. RESEARCH NEEDS

- Research should be conducted on control methods and the most up-to-date and effective information on how to control invasive plants. It is difficult to get an herbicide or a method that is selective enough to kill invasives, but not native plants.

- Restoration methods after applying herbicides are needed.

- The effectiveness of installing a rinsing station at lakes, the costs and benefits of installing stations, and how to effectively design stations need to be research.

- Research on the impacts of recreational boat flow and traffic is needed.

- An AIS rapid response plan needs to be developed to address new or expanding AIS species.

- Ohio's State Management Plan for AIS needs to be revised to incorporate up-to-date information.

A.37. SUMMARY OF AQUATIC INVASIVE SPECIES MANAGEMENT IN OKLAHOMA

A.37.1. AQUATIC INVASIVE SPECIES MANAGEMENT PLAN

Plan under development.

A.37.2. AQUATIC INVASIVE SPECIES PROGRAMS AND ACTIVITIES

- **Sport Fish Restoration Project, Oklahoma Department of Wildlife Conservation (ODWC), Fisheries Division (FD).** The Division conducts monitoring and outreach. Outreach programs include posted signs to inform users of the presence of aquatic nuisance species such as white perch, zebra mussels, and hydrilla and explain how to avoid moving them from one body of water to the next. The Division also conducts native aquatic vegetation introductions in several reservoirs to improve nursery habitat for juvenile sport fishes.

- **Oklahoma Golden Alga Response Team, ODWC FD.** The Division is working to devise efficient and effective plans to respond to golden alga fish kills, as well as proactive solutions to potential golden alga blooms.

- **Spring Creek Lakes Alligatorweed Biocontrol Program, ODWC FD.** The Division is conducting biocontrol through the release of an alligatorweed flea beetle.

- **Aquatic Vegetation Control, ODWC FD.** The Division used grass carp to control vegetation in some of the state fishing lakes, as well as in some state fish hatcheries. Some municipalities have also used grass carp to control vegetation in city-water-supply lakes. A recently formed multi-agency Hydrilla Task Force will address recent infestations of the exotic weed in three reservoirs.

- **Oklahoma Zebra Mussel Task Force.** The multi-agency team, which includes the ODWC FD, shares information on agency activities related to zebra mussel monitoring. The Division has developed Hazard Analysis and Critical Control Point (HAACP) plans to avoid spreading aquatic invasive species through hatchery and management activities.

A.37.3. CLIMATE CHANGE CONCERNS

- Zebra mussels may be able to inhabit warmer environments successfully.
- Temperature changes may contribute to golden alga blooms.

A.37.4. CLIMATE CHANGE ACTIONS

(None reported.)

A.37.5. RESEARCH ACTIVITIES AND INFORMATION USED

- After white perch were discovered in the Kaw Reservoir in 2000, the Division began a 4-year research project to investigate the problem. Results showed white perch never reached high levels. Although reproductive success was high each year, recruitment of to-age-1 individuals was low. No adverse effects on other native fish species in Kaw Reservoir were identified during the research period.

- The ODWC has provided funding to Oklahoma State University for the following research projects:

  o Determining the impacts of zebra mussels on biodiversity on selected rivers within the Tallgrass Prairie Ecoregion.

  o Monitoring water quality parameters and alga abundance at Lake Texoma to determine triggers for golden alga blooms.

  o Determining the toxicity of golden alga toxins to selected species of Lake Texoma fishes and what physical and biological parameters trigger toxin production.

A.37.6. RESEARCH NEEDS

- Controls for white perch, zebra mussels, and hydrilla are needed.

- Research on golden alga (prediction and eradication) is needed.

- Restoration is needed, including introduction of native aquatic plants in ponds, lakes, and reservoirs that contain a variety of herbivores (carp, turtles), fluctuating water levels, and turbidity issues.

A.38. SUMMARY OF AQUATIC INVASIVE SPECIES MANAGEMENT IN OREGON

A.38.1. AQUATIC INVASIVE SPECIES MANAGEMENT PLAN

Oregon's Aquatic Nuisance Species Management Plan was published in 2001 (see Appendix B, State Aquatic Invasive Species Management Plan Summaries for a general description of the Plan).

A.38.2. AQUATIC INVASIVE SPECIES PROGRAMS AND ACTIVITIES

- **Invasive Species Council.** The Council focuses on preventing the new introductions of species, outreach and education programs, and coordinating all agencies involved in aquatic species management.

- **Oregon Clean Safe Boating Program, Oregon State Marine Board (OSMB).** The OSMB conducts a clean boating and invasive species awareness campaign. It develops brochures, illustrated panels, and demos of specimens for trade show exhibits. It maintains a Website and produces a newsletter that goes to every registered boater in the state. As of December 2006, OSMB is working on a Clean Marina Program that will develop an incentive to encourage good housekeeping, conduct training for law enforcement, and create best management practices for facilities development.

- **Lake Lytle Milfoil Control Project, Oregon State Weed Board (OSWB).** The OSWB developed the *Integrated Aquatic Vegetation Management Plan for Lake Lytle*. The plan's first year included application of aquatic herbicide Sonar, as well as pre- and post-treatment vegetation sampling, quality sampling, and an information/education component.

- **Noxious Weed Program, Oregon Department of Agriculture.**

A.38.3. CLIMATE CHANGE CONCERNS

- Climate change raises the question of whether working on aquatic invasive species is fruitless. Species will move because of changes in climate, which may be part of a natural cycle. Certain species in Oregon are more prevalent or less prevalent with El Niño and La Niña patterns, for example.

A.38.4. CLIMATE CHANGE ACTIONS
(None reported.)

A.38.5. RESEARCH ACTIVITIES AND INFORMATION USED
(None reported.)

A.38.6. RESEARCH NEEDS

- More demographic information, e.g., the 100th Meridian Program is doing surveys on the mobility of boaters to determine where to erect signs, is needed

- Scientific information on how to best sanitize boats is needed.

A.39. SUMMARY OF AQUATIC INVASIVE SPECIES MANAGEMENT IN PENNSYLVANIA

A.39.1. AQUATIC INVASIVE SPECIES MANAGEMENT PLAN

Pennsylvania's Aquatic Invasive Species (AIS) Management Plan was published in October 2006 (see Appendix B, State Aquatic Invasive Species Management Plan Summaries for a general description of the Plan).

A.39.2. AQUATIC INVASIVE SPECIES PROGRAMS AND ACTIVITIES

- **Stream ReLeaf Program, Pennsylvania Department of Environmental Protection (PA DEP).** PA DEP holds riparian plant identification classes for staff from regional offices, county conservation districts, and watershed groups. The classes cover the importance of riparian buffers and restoration projects, as well as biodiversity and native and invasive plants.

- **Delaware River Invasive Plant Partnership, States of Delaware, New Jersey, New York, and Pennsylvania.**

- **Zebra Mussel Control, private water suppliers.** The water suppliers apply chemicals, like chlorine, to intake screens on public water supplies to control zebra mussels.

- **Pennsylvania Sea Grant.** Sea Grant has conducted Hazard Analysis and Critical Control Point training for state and federal agencies and developed outreach materials on specific AIS, including materials for AIS prevention among boaters.

- **Zebra Mussel Monitoring, PA DEP.** PA DEP is tracking the distribution and spread of zebra mussels in the Great Lakes region. The agency originally set up ~170 monitoring stations across the state and alerts contacts for adjacent water bodies when there is a new discovery.

- **Invasive plant species control, Pennsylvania Department of Conservation and Natural Resources (PA DCNR).** PA DCNR controls invasive plant species on the lands and in the associated waters that it manages with systemic herbicides and mechanical and biocontrols.

- **Pennsylvania Invasive Species Council.** The Council, established by executive order in 2004, advises the Governor on invasive species issues in Pennsylvania. The council is also charged with (1) *Developing and implementing* a comprehensive invasive species management plan for the state; (2) *Providing guidance* on the prevention and control of nonnative invasive species and rapid response to new infestations; and (3) *Facilitating coordination* among federal, regional, state, and local initiatives and organizations engaged in the management of nonnative invasive species. The Council is comprised of seven state agencies and 10 at-large members.

A.39.3. CLIMATE CHANGE CONCERNS
(None reported.)

A.39.4. CLIMATE CHANGE ACTIONS
(None reported.)

A.39.5. RESEARCH ACTIVITIES AND INFORMATION USED

- Pennsylvania Sea Grant has funded the following AIS research projects (more information available at http://pserie.psu.edu/seagrant/research/ais.htm):

  o Round Goby (*Neogobius melanostomus*) Diet, Habitat Preference, and Reproductive Strategies in Presque Isle Bay

  o Population Assessment of Rudd (*Scardinius erythrophthalmus*) in Presque Isle Bay, Lake Erie

  o Distribution of the Invasive Red-Eared Slider Turtle (*Trachemys scripta elegans*) in the Lower Delaware River Basin

  o A Benthic Survey of the Natural Lakes of Northwestern Pennsylvania

  o Effect of Non-Native Mollusk Species on Common Map Turtles, *Graptemys geographica*

  o Impact of the Round Goby (*Neogobius melanostomus*) on Tributary Streams of Lake Erie

  o A Sampling of Presque Isle Bay for the Exotic Cladoceran: *Bythotrephes cederstroemi*

  o Characterization of the Microplanktonic and Microbenthic Communities of Near-Shore Lake Erie

  o Monitoring Zebra Mussel Invasion of Edinboro Lake, Conneauttee Creek, and French Creek

- Pennsylvania Sea Grant and partners conducted a pilot study on the distribution and sensory biology of the flathead catfish in order to develop strategies to prevent its spread.

A.39.6. RESEARCH NEEDS

- Economic impacts of AIS in Pennsylvania need to be examined.
- Species-specific control technologies are needed.

A.40. SUMMARY OF AQUATIC INVASIVE SPECIES MANAGEMENT IN RHODE ISLAND

A.40.1. AQUATIC INVASIVE SPECIES MANAGEMENT PLAN
Plan under development.

A.40.2. AQUATIC INVASIVE SPECIES PROGRAMS AND ACTIVITIES

- **Mute Swan Management Program, Rhode Island Department of Environmental Management (RIDEM), Division of Fish and Wildlife.** The Division identifies nests and destroys eggs by addling or puncturing them during the swan nesting season.

- **Permit reviews for herbicide application, RIDEM Division of Fish and Wildlife, Rhode Island Department of Agriculture.** RIDEM Division of Fish and Wildlife issues permits for landowners wishing to use chemical treatments to exterminate aquatic invasive species on private or public waters.

A.40.3. CLIMATE CHANGE CONCERNS

- Mute swans may expand their range because of climate change.

- Narragansett Bay ecosystem may respond to a warming trend, including changes in nutrient cycling.

A.40.4. CLIMATE CHANGE ACTIONS
(None reported.)

A.40.5. RESEARCH ACTIVITIES AND INFORMATION USED

- Two rapid assessment surveys (2001 and 2003) have taken place through the MIT Sea Grant.

- Several species-specific studies of aquatic invasives in Rhode Island have been conducted and are currently used by the research community. These studies are also used as baseline data for the state management plan.

A.40.6. RESEARCH NEEDS

- Public perception of swan euthanization and methods for public education and outreach to overcome public discontent should be researched.

- Research on swan control methods is needed (e.g., capturing birds during molting season when they cannot fly).

- Better product information and data about the half lives of herbicides and the effect of their residues is needed. It will be necessary to conduct assay tests to better determine the effects of pesticides on water quality.

- Further baseline studies are necessary for the bay ports of Providence, Quonset, and Newport.

- Baseline studies beyond rapid assessment survey floating dock studies, including those that capture information on sub-tidal benthic and rocky intertidal communities, need to be conducted.

A.41. SUMMARY OF AQUATIC INVASIVE SPECIES MANAGEMENT IN SOUTH CAROLINA

A.41.1. AQUATIC INVASIVE SPECIES MANAGEMENT PLANS

No plan available.

A.41.2. AQUATIC INVASIVE SPECIES PROGRAMS AND ACTIVITIES

- **Aquatic Nuisance Species Program, South Carolina Department of Natural Resources (SCDNR), the South Carolina Aquatic Plant Management Council.** SCDNR, in coordination with the Council, develops and implements an annual management plan for the state, which includes identification of problem areas, a management strategy for the problem areas, and a budget. Management strategies include chemical controls, environmental controls (e.g., water draw-down in lakes, nutrient loading), surveys for invasive species, biocontrols, and mechanical harvesting. The annual management plan is submitted for a 30-day public review period in which all comments received are addressed and modifications are made to the plan.

- **Analytical and Biological Services, Santee Cooper (South Carolina Public Service Authority, a quasi-public entity).** Santee Cooper actively surveys for aquatic invasive plants on Lakes Marion and Moultrie. All control operations are approved by and coordinated through the state Aquatic Plant Management Plan. The Water Quality Monitoring Program tests the water for invasive species two or three times a week and conducts aerial aquatic plant surveys of the lake system annually. Control efforts for hydrilla include the stocking of sterile grass carp. For water hyacinth, herbicides are sprayed from a helicopter or airboats as needed. For alligatorweed and water primrose, spot chemical treatments are applied as needed.

A.41.3. CLIMATE CHANGE CONCERNS REPORTED BY STATE PERSONNEL

- Some plant species that are sensitive to cold weather, such as water hyacinth and water lettuce, have started to move north and inland.

A.41.4. CLIMATE CHANGE ACTIONS

(None reported.)

A.41.5. RESEARCH ACTIVITIES AND INFORMATION USED

(None reported.)

A.41.6. RESEARCH NEEDS

- Statewide mapping of the range of invasive species or a "census" of invasive species need to be conducted, so that control programs can map their progress in controlling and eradicating pests.

A.42. SUMMARY OF AQUATIC INVASIVE SPECIES MANAGEMENT IN SOUTH DAKOTA

A.42.1. AQUATIC INVASIVE SPECIES MANAGEMENT PLAN

No plan available.

A.42.2. AQUATIC INVASIVE SPECIES PROGRAMS AND ACTIVITIES

- **South Dakota/Nebraska Purple Loosestrife Management Committee, Wildlife Management Institute, South Dakota Department of Agriculture (SDDA), and counties, federal agencies, local agencies, universities, and other South Dakota and Nebraska state agencies).** The Committee developed a large-scale purple loosestrife biocontrol rearing and redistribution facility and several satellite locations that are being managed by local county weed and pest personnel. Control is conducted using purple loosestrife biocontrol beetles and aerial and ground spraying with Roundup™.

- **Tamarisk Mapping, SDDA Office of Agricultural Services.** The Office of Agricultural Services conducts a mapping project and a cooperative management program for tamarisk control and, where possible, eradication. There is a tamarisk task force for Lake Oahe. The Office has released biocontrol agents and placed Tamarisk on the South Dakota noxious weed list.

- **Western Zebra Mussel Task Force, South Dakota Game Fish and Parks Department (SDGFP).** The Department provides dock signage describing how boaters can prevent spread of zebra mussels and other aquatic exotics and is monitoring Lewis and Clark Lake. Education efforts focus on prevention. Biologists and private citizens sample and monitor for zebra mussels.

- **Western Regional Panel, SDGFP, U.S. Fish and Wildlife Service Regional Fisheries Program.** The Program has carried out a variety of activities: (1) *Hosted* the Missouri River Basin/Lewis and Clark Bicentennial ANS workgroup meeting that discussed information/education and outreach strategies to prevent the introduction and spread of ANS in the Missouri River basin; (2) *Revised bait regulations* in the South Dakota Fishing Handbook to limit the type and amount of bait that may be transported into South Dakota (It is working on regulating the harvest of bait below Gavin's Point Dam on the Missouri River where Asian carp have become well-established.); (3) *Continues work on the installation* of at least two Traveler Information Systems (TIS) along the Missouri River. A TIS station would broadcast a message regarding aquatic nuisance species (ANS) and other topics of interest (boat ramp condition, Lewis and Clark events); (4) *Installs ANS signs* at boat ramps; (5) *Works with an SDGFP information specialist* to send out a mailing packet to all state resident fishing license holders (including information regarding ANS, ANS ID cards, adhesive tape measures with ANS prevention message, etc.); and (6) *Researches and monitors* Asian Carp movement.

- **Influence of an introduced diatom (*Didymosphenia geminata*) and directed control measures on the biological community composition of Rapid Creek, SDGFP.** A study is currently being developed to examine the impact of *Didymosphenia geminata* on benthic and fish community composition of Rapid Creek below Pactola Dam. Research will also study the effects of control measures (localized nutrient enrichments) on *Didymosphenia geminata* distribution and overall stream biological community composition.

A.42.3. CLIMATE CHANGE CONCERNS

- Originally South Dakota did not think tamarisk could survive in warm temperatures, but it seems to be adapting.

- The state's 5-year drought has led to a severe increase in the population of tamarisk. When water shrinks back from the edge of lakes or rivers, tamarisk is able to grow in this habitat.

A.42.4. CLIMATE CHANGE ACTIONS
(None reported.)

A.42.5. RESEARCH ACTIVITIES AND INFORMATION USED
(None reported.)

A.42.6. RESEARCH NEEDS

- More on-the-ground surveying and more plant recognition capability are needed.

- Rise in mussel activity needs to be identified through monitoring efforts.

- Information should be distributed to those who use the state's water bodies. Outreach and education is currently on a project-by-project basis due to lack of capacity.

- The biological impacts of curlyleaf pondweed on lake ecosystems need to be better understood.

- Targeted monitoring for ANS presence in lakes needs to be implemented throughout South Dakota.

- A rapid response strategy for ANS detection and management needs to be developed in South Dakota.

- An overall strategic plan for ANS needs to be developed, extending beyond the responsibilities of SDGFP. This overall plan should incorporate involvement from federal, state, local, and private interests throughout the state.

A.43. SUMMARY OF AQUATIC INVASIVE SPECIES MANAGEMENT IN TENNESSEE

A.43.1. AQUATIC INVASIVE SPECIES MANAGEMENT PLAN

No plan available.

A.43.2. AQUATIC INVASIVE SPECIES PROGRAMS AND ACTIVITIES

- **Aquatic plant management, Nickajack Reservoir, the Tennessee Valley Authority (TVA), Marion County.** Aquatic plants are managed along near-shore areas along developed shorelines, and they are controlled to maintain access lanes to open water. Management is primarily for hydrilla and in accordance with a stakeholder-developed plan that prescribes control methods including the use of herbicides in near-shore areas (with a state permit) and mechanical and manual culling.

- **Aquatic plant management, Chickamauga Reservoir, the TVA and private homeowners.** Aquatic plants (spinyleaf naiad and other species) are managed along near-shore areas along developed shorelines, and they are controlled to maintain access lanes to open water in accordance with a stakeholder-developed plan that prescribes private shoreline property owners to use herbicides in near-shore areas (with a state permit) and TVA to mechanically cull aquatic invasive plants.

- **Monitoring and eradication, Obed Wild and Scenic River.** Authorities monitor for purple loosestrife and eradicate (through removal and chemical control), chemical) as needed. They also monitor for exotic mussels, including zebra mussels.

- **Fish monitoring, University of Tennessee.** The University is collecting fish for a project that involves mapping species communities in rivers and streams across Tennessee, including all non-native or invasive species.

- **Eradication and restoration, Warner Parks (Metro Park System), Tennessee Department of Agriculture, Cumberland River Compact Association, Harpeth River Watershed Association, Natural Resources Conservation Service, Friends of Warner Park, and the Eagle Scouts.** The group is conducting a restoration project along Harpers River, where heavy traffic causes riparian buffer damage. Activities include rebuilding the buffer, stopping mowing, fencing off the area, removing invasives, and transplanting native species. Monitoring, removing, and replanting will likely continue as needed.

- **Monitoring and control, Metro Park System, Belmont University.** Monitoring and manual removal of garlic mustard plant is being conducted around the Shelby Bottoms section of the Cumberland River.

- **Species removal and restoration, Great Smoky Mountains National Park, National Park Service, U.S. Environmental Protection Agency, Tennessee Wildlife Resources Agency, North Carolina Wildlife Resources Commission, Tennessee Department of Wildlife and Conservation, Trout Unlimited National, Federation of Fly Fishermen, and others.** Rainbow trout populations in select stream segments above natural barriers are being removed with the fish toxicant antimycin or using backpack electrofishing. Monitoring continues for 1–2 years and then, if rainbow trout have not returned, brook trout (native) are reintroduced.

- **Eradication, Big South Fork National Recreation Area.** Riparian invasive plants are treated chemically.

- **Eradication Program, Oak Ridge National Laboratory (ORNL).** The ORNL manages non-native invasive plants in the riparian zones of streams within the Oak Ridge Reservation. Control methods include applying various herbicides, cutting, and mowing. Target species include privet, autumn olive, kudzu, lespedeza, princess tree, mimosa, and tree of heaven. ORNL also monitors fish and aquatic invertebrates in the streams, recording abundance and distribution of native and non-native species. The National Park Service and The Nature Conservancy conducted a complete vascular plant inventory at the park, which formed the basis of which species should be targeted for removal. The Tennessee Exotic Pest Plant Management Manual was also consulted.

A.43.3. CLIMATE CHANGE CONCERNS

(None reported.)

A.43.4. CLIMATE CHANGE ACTIONS

(None reported.)

A.43.5. RESEARCH ACTIVITIES AND INFORMATION USED

- An investigation is being conducted on the effects that the western mosquito fish is having on efforts to reintroduce the barrens top minnow in Western Tennessee. Researchers want to determine the relationship between the two species and what they can do to alleviate some of the problems.

A.43.6. RESEARCH NEEDS

- Research on the ozone effects on Barrens top minnow is needed.

- Research how the hemlock wooly adelgid affects native hemlock and fish populations should be continued.

- The state should assist the U.S. Environmental Protection Agency in its effort to re-register antimycin.

- More information on burning as a control method needs to be acquired.

- More information on interactions between chemicals and other native animals/plants in the area needs to be collected.

A.44. SUMMARY OF AQUATIC INVASIVE SPECIES MANAGEMENT IN TEXAS

A.44.1.   AQUATIC INVASIVE SPECIES MANAGEMENT PLAN

Plan under development.

A.44.2.   AQUATIC INVASIVE SPECIES PROGRAMS AND ACTIVITIES

- **Golden Alga Task Force/Kills and Spills Team, Texas Parks and Wildlife Department (TPWD), Inland Fisheries.** The program responds to fish and wildlife kills and pollution incidents, minimizes environmental degradation, conducts compensation, repair, and restoration for environmental damage, and monitors golden alga levels. The program also provides education on the relationship between water quality, habitat, and living organisms.

- **Aquatic Habitat Enhancement Program, Nuisance Vegetation Control, Texas TPWD, Inland Fisheries.** This program focuses on the control of aquatic vegetation that affects the health and recreational use of TPWD-managed fish and wildlife resources.

- **Texas Invasive Species Coordinating Committee, formed as of December 2006, involving eight state agencies.** The goal of this multi-agency committee is to facilitate cooperation among state agencies and to help prevent, control, and manage invasive species.

A.44.3.   CLIMATE CHANGE CONCERNS

- Warmer winters and lack of freezing winter temperatures may contribute to the persistence and spread of introduced invasive aquatic vegetation species such as water hyacinth, giant salvinia, and common salvinia.

A.44.4.   CLIMATE CHANGE ACTIONS

(None reported.)

A.44.5.   RESEARCH ACTIVITIES AND INFORMATION USED

- Research golden alga is being conducted, including

  o using clay treatments to control golden alga blooms;

  o determining the economic impacts of golden alga fish kills on the Possum Kingdom area;

- monitoring water quality during a bloom on Lake Whitney, examining genetics and developing diagnostic determinations for events using genetic markers;

- using barley straw to control outbreaks; and

- determining nutrient and water quality parameters that influence bloom and toxin formation.

- Research is being conducted on control of giant salvinia (*Salvinia molesta)* through the use of *Cyrtobagous salviniae*, a biocontrol agent. Giant salvinia propagation, reproduction, and dispersal rates will be examined, as will the potential of *Cyrtobagous salviniae* in long-term control and management.

- Research is being conducted to evaluate the duration an extended summer and fall have on drawdown in BA Steinhagen Reservoir in East Texas and on how this drawdown affects aquatic invasive vegetation.

- Grass carp is being tracked in Lake Austin, Lake Conroe, and the Rio Grande.

- Giant salvinia weevil is being evaluated in the Toledo Bend Reservoir and Lake Conroe.

- Impacts of *Arundo donax* on fishes of the Rio Grande are being evaluated.

- Research is taking place on applesnail (*Pomacea spp.*), including the geographic range of the applesnail invasion in Southeast Texas and its taxonomy and ecology.

A.44.6. RESEARCH NEEDS

- Research needs to be conducted on golden alga control techniques and toxin production, analytical methods to define toxins; frequency and regularity of golden alga's occurrence; and effects of golden alga on the recruitment of fish, soil conditions, runoff, and nutrient loading.

- Natural algaecide compounds need to be tested.

- Transferable methods to estimate the economic impacts of fish kill events on communities need to be found.

- Research on the impact of drought on water hyacinth and hydrilla is needed.

- The physiology and pathways of the grass carp and how the grass carp relates to hydrilla control should be studied.

- Evapotranspiration rates for *Arundo donax* and salt cedar, as compared to native vegetation rates, should be examined.

- The impacts of *Arundo donax* infestations on channelization and stream fishes need to be researched.

- Remote sensing and acreage estimations need to be performed for *Arundo donax*, salvinia, water hyacinth, waterlettuce, saltcedar, and other aquatic invasive vegetation.

- The impacts of Eurasian watermilfoil weevils on *Myriophyllum spicatum* in the Rio Grande need to be evaluated.

- Research on the applesnail infestations of Texas rice crops and native riparian vegetation needs to be conducted.

- Chinese tallow control efforts need to be evaluated.

- The impacts of grass carp on the Galveston Bay Ecosystem; the impacts of *Arundo donax* wasps on giant reed populations; and the conditions for hydrilla expansion should be researched.

- Aquatic invasive species (AIS) should be monitored and tracked in freshwater and estuarine systems to facilitate early detection and rapid response.

- Research needs to be conducted on the ecological, social, and economic impacts of emerging AIS in Texas's coastal watersheds, bays, and estuaries.

A.45. SUMMARY OF AQUATIC INVASIVE SPECIES MANAGEMENT IN UTAH

A.45.1. AQUATIC INVASIVE SPECIES MANAGEMENT PLAN

No plan available.

A.45.2. AQUATIC INVASIVE SPECIES PROGRAMS AND ACTIVITIES

- **Monitoring Program, Utah Department of Natural Resources, Division of Parks and Recreation and Division of Wildlife Resources.** The Program inventories 15–20 waters annually for zebra mussels, educates drivers of vehicles from areas of known zebra mussel infestations, encourages boat washing at the Division's expense, and inspects 10% of boats for infestations. The Program also posts public alert signs at major recreational waters, includes aquatic nuisance species (ANS) information inserts in boat re-licensing packets, and prints and distributes ANS brochures to major boating information centers, boat dealers, and sporting goods outlets. New Zealand mud snail brochures have also been printed. The Program also surveys docks and buoys at the end of each summer season for signs of mussels, snails, and Eurasian watermilfoil. Finally, the Program maintains kiosks and posts information about anglers' responsibilities in keeping boats clean.

- **Recovery Program, State of Colorado, State of Utah, U.S. Fish and Wildlife Service, Colorado State University.** The Program conducts research, removal, and relocation to area fishing ponds wherever appropriate and practical, as well as euthanization of invasive fish.

- **Biosecurity Measures, Utah Division of Wildlife Resources.** Biosecurity measures have been standardized for all aquatic personnel within the Division who conduct surveys and sampling, so as to prevent the movement of ANS between habitats.

- **Hatchery Monitoring, Utah Division of Wildlife Resources.** In addition to monitoring public and private waters for ANS, the Division has been actively engaged in monitoring state-owned hatcheries for ANS. Whirling disease is a particular concern as there have been three infected hatcheries. Mammoth Creek Hatchery has been reconstructed and disinfected. Reconstruction on Midway Hatchery began in the winter of 2007 and is scheduled to be completed in June 2008. The Division also has submitted a proposal to construct sand filtration and UV exposure systems for water sources that feed into the Springville Hatchery.

- **New Zealand Mud Snails Cooperative Studies, Utah State University.** An on-going study at Utah State University is focusing on interactions between the New Zealand mud snail and trout in the Green River. Recent reports indicate that trout may help spread the snail.

A.45.3. CLIMATE CHANGE CONCERNS

(None reported.)

A.45.4. CLIMATE CHANGE ACTIONS

(None reported.)

A.45.5. RESEARCH ACTIVITIES AND INFORMATION USED

(None reported.)

A.45.6. RESEARCH NEEDS

- Research is needed on the New Zealand mud snail and on ways to prevent the spread of the zebra mussel.

A.46. SUMMARY OF AQUATIC INVASIVE SPECIES MANAGEMENT IN VERMONT

A.46.1. AQUATIC INVASIVE SPECIES MANAGEMENT PLAN
No plan available.

A.46.2. AQUATIC INVASIVE SPECIES PROGRAMS AND ACTIVITIES

- **Aquatic Nuisance Control Program, Vermont Department of Environmental Conservation (VT DEC).** The program's goal is "to prevent or reduce the environmental and socio-economic impacts of nuisance (primarily non-native) aquatic plant and animal species." The program's seven sub-programs include the following:

  - The Aquatic Nuisance Species (ANS) Watchers Program—includes training for interested volunteers to assist in early detection (species identification, lake searches, and communication of the status of nuisance species)

  - The Purple Loosestrife Biocontrol Program—includes on-the-ground management and control activities such as introducing leaf-eating beetles (*Galerucella spp.*), a biocontrol; selecting sites; obtaining landowner permission; monitoring; increasing public knowledge; raising and releasing beetles; and compiling and summarizing activities and findings

  - The Water Chestnut Management Program—includes on-the-ground management and control activities such as mechanical harvesting, manual culling, surveying, education, and outreach

  - The Eurasian Watermilfoil (EWM) Biocontrol Program—includes on-the-ground management and control activities such as weevil introductions and augmentations as a biocontrol agent for Eurasian watermilfoil

  - The Grant-in-Aid Program—provides financial assistance to municipalities and agencies for the control of EWM, as well as ANS spread prevention, mechanical control of nuisance native plant populations, and management of purple loosestrife

  - The permitting program for mechanical and chemical control of invasive species

  - Spread prevention

- **Alewife Monitoring, Vermont Department of Fish and Wildlife.** Since the discovery of this aquatic invasive fish species in Lake Champlain in 2004, monitoring activities on Lake St. Catherine and downstream waters has ceased as has research on control or eradication measures. Instead, activities now focus on monitoring the spread and increase of alewife in Lake Champlain. Current emphasis is on aquatic invasive species public education and outreach.

- **Regulatory Development, VT Department of Fish and Wildlife, ANS Team**. The VT Department of Fish and Wildlife ANS Team works to create new rules and regulations or amendments to existing rules and regulations that work to prevent or reduce the risk of aquatic exotic species introductions. In the past, existing rules pertaining to the baitfish industry were revised and included the creation of a permitting program for the importation, harvesting, and sale of baitfish. A baitfish identification booklet was also published. Currently, two regulations are being drafted—one pertains to general fish importation where the intent is to stock the fish and the second adopts prohibited, restricted, and unrestricted fish species lists and a permitting requirement on the importation of fish species regardless of intent. This rule will also pertain to the aquarium trade.

- **Public education and outreach, VT Department of Fish and Wildlife.** Efforts to increase public awareness of exotic species issues, concerns, and risks are ongoing. Activities include work with the baitfish industry to write and adopt Hazard Analysis and Critical Control Point planning protocols into their daily operation.

- **Sea Lamprey Control Program, VT Department of Fish and Wildlife, New York Department of Conservation, and U.S. Fish and Wildlife Service.** The program uses a variety of methods to control sea lamprey, including trapping adults in smaller spawning streams, constructing and maintaining barriers on certain streams to prevent sea lampreys from reaching spawning areas, and periodically using chemical lampricides to kill young sea lampreys in larger streams and rivers.

- **Lake Champlain Zebra Mussel Monitoring Program, VT DEC and Lake Champlain Basin Program.** Project activities include: (1) *Monitoring* the distribution and abundance of zebra mussel larvae, juveniles, and adults; (2) *Determining the occurrence* of new colonization in Lake Champlain, tributaries, and inland lakes and incorporating this information into a database; (3) *Determining appropriate management responses* and assessing the effectiveness of spread prevention or control measures; (4) *Informing the public, water treatment facility operators, and marina managers* about zebra mussels so that appropriate spread prevention and control measures are taken; (5) *Providing technical assistance* on the design and operation of zebra mussel monitoring programs; (6) *Documenting water quality parameters* pertinent to zebra mussel survival; (7) *Producing a report* that documents the findings of the Lake Champlain Zebra Mussel Monitoring Program; and (8) *Maintaining* the Lake Champlain Zebra Mussel Monitoring Program Website.

- **Lake Champlain Basin ANS Management Plan, VT DEC, New York Department of Environmental Conservation, in cooperation with state and federal agencies, regional bodies, and nongovernmental organizations.** The plan focuses on facilitating the coordination of ANS management efforts, providing opportunities for federal cost sharing, and implementation.

A.46.3. CLIMATE CHANGE CONCERNS

(None reported.)

A.46.4. CLIMATE CHANGE ACTIONS

(None reported.)

A.46.5. RESEARCH ACTIVITIES AND INFORMATION USED

- Research continues to provide new non-chemical control methods to reduce reliance on lampricides.

A.46.6. RESEARCH NEEDS

- The current distribution of specific ANS and the impacts they have on ecosystems and native species needs to be researched.

- The economic impacts of ANS need to be determined.

- The reason Phragmites have appeared where beetles have reduced the presence of purple loosestrife needs to be identified.

- The impacts of ANS in other states and effectiveness of control programs needs to be researched.

- Monitoring needs to take place for new ANS.

- Sea lamprey control technology needs to be developed.

- The effectiveness of filtering to control plankton populations and of using pheromones to lure lampreys needs to be evaluated.

- The densities of zebra mussels throughout their life stages need to be monitored.

A.47. SUMMARY OF AQUATIC INVASIVE SPECIES MANAGEMENT IN VIRGINIA

A.47.1. AQUATIC INVASIVE SPECIES MANAGEMENT PLAN

Virginia's Invasive Species Management Plan was published in 2005 (see Appendix B, State Aquatic Invasive Species Management Plan Summaries for a general description of the Plan).

A.47.2. AQUATIC INVASIVE SPECIES PROGRAMS AND ACTIVITIES

- **Cooperative Project, Virginia Department of Conservation and Recreation (VA DCR) and Virginia Native Plant Society.** The project seeks to identify alien plant species that have the potential to become invasive; document threats; coordinate with other agencies and organizations to identify mutual concerns; develop solutions; and develop and implement sound practices for the control of invasive alien plants in natural areas.

- **Snakehead Sampling (monitoring program), Virginia Department of Game and Inland Fisheries (VDGIF).** The program involves intensive sampling in one to two small creeks or streams.

- **Legislation, Virginia Legislature.** The Aquatic Invasive Species (AIS) Act increased criminal and civil penalties and gave the Board authority to add additional AIS to the states list of Nonindigenous Aquatic Nuisance Species (ANS). The law applies to any species with the potential to cause statewide impact.

- **Phragmites Control, VA DCR, in conjunction with VDGIF, U.S. Fish and Wildlife Service, The Nature Conservancy, National Park Service, U.S. Department of Defense, Virginia Institute of Marine Science.** VA DCR has mapped the distribution of Phragmites and targets certain areas for control efforts, which include the aerial application of herbicides.

- **Chinese Mitten Crab, VDGIF.** The VDGIF has recommended that the Chinese mitten crab be added to the states list of Nonindigenous ANS.

A.47.3. CLIMATE CHANGE CONCERNS
(None reported.)

A.47.4. CLIMATE CHANGE ACTIONS
(None reported.)

A.47.5. RESEARCH ACTIVITIES AND INFORMATION USED

- Research on snakehead, including identifying sampling areas, testing sampling methods, and studying population genetics, is taking place.

- Phragmites distribution is being mapped.

A.47.6. RESEARCH NEEDS

- Pathways and incentives (e.g., how people introduce invasive species) need to be researched in order to educate the public and influence behavior.

- Research needs to be conducted on natural diseases or parasites for the snakehead, as well as methods to capture, control, and/or eliminate the species.

- Researching is needs on how different wildlife use Phragmites.

A.48. SUMMARY OF AQUATIC INVASIVE SPECIES MANAGEMENT IN WASHINGTON

A.48.1. AQUATIC INVASIVE SPECIES MANAGEMENT PLAN

Washington's Aquatic Nuisance Species (ANS) Management Plan was published in October 2001 (see Appendix B, State Aquatic Invasive Species Management Plan Summaries for a general description of the Plan).

A.48.2. AQUATIC INVASIVE SPECIES PROGRAMS AND ACTIVITIES

- **Aquatic Weeds Program, Washington Department of Ecology (Ecology).** The Department provides education, technical assistance, and financial assistance to governments and local lake groups to help them manage the problems caused by invasive non-native freshwater plants. The Department offers grants as "seed" money to initiate freshwater invasive plant species eradication and control projects. Several eradication and control strategies are used, including: hand pulling and bottom barrier installation, aquatic herbicide treatment (2,4-D, fluridone, triclopyr, imazapyr, glyphosate, endothall, diquat), triploid grass carp, diver dredging, harvesting, rotovation, and water level drawdown. Eurasian watermilfoil (EWM), Brazilian elodea, hydrilla, fragrant water lily, yellow flag iris, purple loosestrife, and many other state-listed noxious weeds are eligible for grant-funded projects. As a result of this program, EWM has been eradicated from seven water bodies and many lake groups are keeping milfoil at such low populations that it no longer is posing a threat to recreation and the environment. Ecology is also funding research into the impacts of aquatic herbicides on salmonids (University of Washington), conducting research on "test" lakes after herbicide treatment, and has an ongoing project on biocontrol for EWM (weevils).

- **Prevention Program, Washington Department of Fish and Wildlife (WDFW).** The Program focuses on prevention activities for (1) ballast water, (2) recreational watercraft, and (3) aquatic plant and animal suppliers. The *Recreational Watercraft Program* (Bill 5679) puts a fee on recreational boats. The *Aquatic Plant and Animal Suppliers Program* classifies species into three categories: Prohibited, Regulated, and Unregulated. Activities include sending enforcement officers to inspect pet stores and issue tickets to regulate the release of invasive species and regulating the importation of prohibited species. Washington has list of aquatic invasive species (AIS) that cannot be sold.

- **Control programs, WDFW.** This program focuses on controlling and eradicating invasive tunicates found in several locations around Puget Sound.

- **Early Detection and Rapid Response Program (EDRR), WDFW.** An EDRR Plan has been developed by the ANS Committee. A Memorandum of Agreement (MOA) is currently being drafted between all the natural resource agencies in the state that will be implementing the program. In the case of new species introduction, the MOA will designate a lead agency, the funding source, and the process for managing the new species.

- **Invasive plant control programs, Washington Department of Agriculture.** The Department leads the state's effort to monitor for and eradicate invasive Spartina infestations. The WDFW and Department of Natural Resources also participate in this program. The Program also monitors other invasive plants including purple loosestrife and various non-native invasions of knotweed. The Department also controls the introduction and spread of invasive plants and disease organisms through its quarantine program.

- **State Noxious Weed Control Board.** The Board lists state noxious weeds and works with local weed boards and landowners to control and eradicate invasive aquatic plants infesting private property.

- **Puget Sound Action Team**. The team's staff coordinates and supports a number of activities, including staffing the state Ballast Water Committee, and coordinating the state's response to eradicate invasive tunicates recently found in Puget Sound. In 2006, the Governor and the Legislature provided emergency and supplemental funds to eradicate invasive non-native tunicates. In addition, the Action Team and its advisors on the Puget Sound Council develop a two-year plan and budget to protect and restore Puget Sound, including actions to prevent and control invasive aquatic plants and animals. The plan and budget became part of the Governor's budget to fund activities in the Puget Sound basin.

- **Invasive Species Council.** The 2006 Legislature created this policy level Council to coordinate among state agencies on aquatic and terrestrial invasive species issues. The Office of the Interagency Committee staffs this Council. The Council will prepare a long range strategy for managing invasive species in the state.

A.48.3. CLIMATE CHANGE CONCERNS

- Climate change will likely expand the range of some of AIS.

A.48.4. CLIMATE CHANGE ACTIONS
(None reported.)

A.48.5. RESEARCH ACTIVITIES AND INFORMATION USED

- Ecology is funding the University of Washington to conduct research into the sub-lethal impacts of aquatic herbicides on salmonids.

- Washington State University is conducting herbicide field trials for parrotfeather, yellow flag iris, and hairy willow-herb.

A.48.6. RESEARCH NEEDS

- Information on the types of legislation that may be enacted and on possible funding sources. For example, a state that wants to take a pathway approach for recreational watercraft could benefit from a list of programmatic approaches and a list/summary of state laws, so that states can understand their options.

A.49. SUMMARY OF AQUATIC INVASIVE SPECIES MANAGEMENT IN WEST VIRGINIA

A.49.1. AQUATIC INVASIVE SPECIES MANAGEMENT PLAN

No plan available.

A.49.2. AQUATIC INVASIVE SPECIES PROGRAMS AND ACTIVITIES

- **Monitoring and Control, West Virginia Department of Agriculture (WVDA).** The department surveys and maps hydrilla in selected locations and monitors a beetle released to combat the hemlock wooly adelgid.

- **Control, U.S. Department of Agriculture (USDA) Forest Service, WVDA, and The Nature Conservancy (with grant funds from West Virginia Advisory Board).** The program focuses on control of non-native, invasive species.

- **Appalachian Highlands Invasive Species Project, The Mountain Institute (with grant funds from West Virginia Advisory Board).** The project includes research, education, and a demonstration site to develop control methods that may then be used to grow native plants and restore the area.

- **Monitoring, Control, and Eradication, U.S. Fish and Wildlife Service (USFWS), USDA Natural Resources Conservation Service, West Virginia Department of Forestry, West Virginia Department of Environmental Protection, West Virginia Division of Natural Resources (WV DNR), as well as county and city councils, local garden clubs, and volunteers.** The program identifies and monitors species, educates volunteers, and manages and eradicates purple loosestrife by spraying chemical herbicides.

- **Control and Monitoring, WV DNR, USDA, and various states.** The program breeds and releases Garacella Beetles, which act as a biocontrol for purple loosestrife. Data on breeding, release, plant counts, and spread is collected bi-annually.

- **Monitoring, WV DNR (with grants from the USDA Cooperative Annual Pest Survey).** The program includes general monitoring and weed surveys of pest plants across the state. Field scouts are trained to search for the invasive species, which are then mapped.

- **Monitoring, USFWS and WV DNR.** The program conducts quantitative monitoring for zebra mussels and sampling to estimate biomass and populations.

- **Monitoring and Research, WV DNR.** The program maps the distribution of invasive crawfish in state rivers.

- **Regulation, WV DNR.** The agency requires a permit to stock triploid grass carp in private ponds and any warm water species of fish into public waters.

A.49.3. CLIMATE CHANGE CONCERNS

- General concerns exist about the effects of climate change on species.

A.49.4. CLIMATE CHANGE ACTIONS
(None reported.)

A.49.5. RESEARCH ACTIVITIES AND INFORMATION USED

- Mapping and monitoring are being conducted.

A.49.6. RESEARCH NEEDS

- Further development of control and eradication methods is needed.

- More specific information is needed on the distribution of hydrilla in the state.

- Information needs to be collected on invasive plants (mile-a-minute, Japanese knotweed) and biocontrols.

- More effective plant mapping needs to be implemented.

- Agencies should cooperate more to pool information more effectively.

- Comprehensive ways to determine if a plant is invasive need to be identified.

A.50. SUMMARY OF AQUATIC INVASIVE SPECIES MANAGEMENT IN WISCONSIN

A.50.1. AQUATIC INVASIVE SPECIES MANAGEMENT PLAN

Wisconsin's Comprehensive Management Plan to Prevent Further Introductions and Control Existing Populations of Nonindigenous Aquatic Nuisance Species was published in 2003 (see Appendix B, State Aquatic Invasive Species Management Plan Summaries for a general description of the Plan).

A.50.2. AQUATIC INVASIVE SPECIES PROGRAMS AND ACTIVITIES

- **Aquatic Plant Management Program, Wisconsin Department of Natural Resources (WDNR).** The program seeks to control efforts for Eurasian watermilfoil (EWM) and curlyleaf pondweed through weed harvesting or spot chemical treatment, as well as some biocontrol for EWM.

- **Aquatic Invasive Species (AIS) Program**, WDNR. The program conducts the following activities: (1) *Watercraft Inspection*, including the dissemination of information to anglers and boaters that identifies AIS and what precautions to take, visual inspection and demonstration of the proper steps to clean boats and equipment, and the installation of signs informing boaters of infestation status, state law, and steps to prevent spreading aquatic invasives; (2) *Monitoring* for zebra mussels (including collection of samples for veliger analyses and deployment of substrate samples), EWM (including inspection of watercraft or shorelines for invasive plants), spiny waterfleas, rusty crayfish, and curlyleaf pondweed; (3) *Clean Boats, Clean Waters Volunteer Program* (in cooperation with the University of Wisconsin-Extension and the Wisconsin Association of Lakes), which offers training on how to organize a watercraft inspection program, how to inspect boats and equipment, and how to interact with the public and encourages volunteers to help monitor for aquatic invasives; (4) *Purple Loosestrife Biological Control* (in cooperation with the University of Wisconsin-Extension), which is a citizen-based project that emphasizes the use of two beetle species for biocontrol, in combination with traditional methods, and conducts some mechanical harvesting and monitoring of impact; and (5) *Information and Education* (in cooperation with the University of Wisconsin-Extension and Wisconsin Sea Grant), with a focus on working with resource professionals and citizens statewide to teach water users the steps to prevent transporting aquatic invasives, as well as addressing aquarium pet release and water gardening (educational tools include brochures and publications, watch cards and wild cards, public service announcements, and displays at parks, sport shows, state fair, conventions and symposiums).

- **Invasive Species Awareness Month (June), WDNR in cooperation with various nongovernmental organizations.** Workshops, field trips, lectures, and work parties are held statewide in June as part of Invasive Species Awareness Month for Wisconsin. Activities include AIS displays with handouts and experts on site.

- **Citizen Lake Monitoring Network (formerly Self-Help Citizen Lake Monitoring), WDNR, University of Wisconsin-Extension and Wisconsin Lakes Partnership.** With over 1,200 trained citizen volunteers statewide, project goals are (1) to collect high quality data, (2) to educate and empower volunteers, and (3) to share data and knowledge. Volunteers learn to identify exotics and are the eyes for water biologists in helping to monitor the state's 15,081 lakes. Volunteers monitor for EWM, curlyleaf pondweed, purple loosestrife, rusty crayfish, zebra mussels, and waterfleas.

- **Wisconsin Invasive Plants Reporting and Prevention Project, WDNR, University of Wisconsin, Wisconsin State Herbarium, and others.** The initiative focuses on early detection and rapid response. Special public recognition is given to those who are among the first to find new invasive species in Wisconsin. In addition, collected specimens become part of the permanent collection of the Wisconsin State Herbarium.

- **AIS Grants, WDNR.** This program awards grants to local municipalities, on a 50% cost-share basis, for AIS control, including prevention, eradication of pioneer populations, planning and education, and restoration.

A.50.3. CLIMATE CHANGE CONCERNS

- Over the next century many species found in northern Illinois could survive in Wisconsin. New species may take over with any shift in climate, particularly if native species cannot adapt. Fish are especially vulnerable. For example, trout have a narrow tolerance range for temperature; if the temperature in headwater streams rises by three to five degrees, those trout may be threatened and niches may open up for AIS such as Asian carp.

A.50.4. CLIMATE CHANGE ACTIONS

- A professor at the Center for Limnology has been funded to study climate change impacts.

A.50.5. RESEARCH ACTIVITIES AND INFORMATION USED

- Studies have been conducted on biocontrol (native beetles) for EWM.

- Pilot tests have been conducted on a dozen or more lakes to lessen the impact from AIS.

- Database management captures all monitoring data and watercraft inspection. Research on building a system is ongoing.

- Model predictions are being conducted to determine which lakes are more vulnerable to AIS.

A.50.6. RESEARCH NEEDS

- Research is needed on hybrid watermilfoil. WDNR has discovered a hybrid of EWM (a cross between Eurasian and northern milfoil) and associated implications regarding control methods. The effects of chemicals on the hybrid are not fully understood. Research on the physical identification of the hybrid strains would also be useful. Because hybrids closely resemble EWM, currently the only way to identify is through genotyping, which is very expensive. Research on the origin of the hybrid would also assist in understanding how it is generated.

- Research is needed on infestation. Determining how to predict which waters would be most vulnerable to infestations by AIS would help focus monitoring efforts. For instance, low calcium and Ph levels can hinder establishment and reproduction of zebra mussels.

- Research is needed on successful rapid response methods (i.e., trapping out crayfish to allow native species to rebound, control of rainbow smelt by dumping in more walleyes, and introducing bass to control crayfish).

A.51. SUMMARY OF AQUATIC INVASIVE SPECIES MANAGEMENT IN WYOMING

A.51.1. AQUATIC INVASIVE SPECIES MANAGEMENT PLAN

No plan available.

A.51.2. AQUATIC INVASIVE SPECIES PROGRAMS AND ACTIVITIES

- **Evaluation of the Efficiency and Efficacy of Non-Native Fish Eradication and Exclusion Techniques for Native Fish Restoration (2004–2005), Montana Fish, Wildlife, and Parks, Wyoming Game and Fish Department, U.S. Fish and Wildlife Service, Yellowstone National Park, Wild Fish Habitat Initiative.** The project entails construction of fish barriers to prevent passage of non-native trout (particularly Brook Trout), as well as chemical treatments using the pesticides Animiasin and Rotenone.

A.51.3. CLIMATE CHANGE CONCERNS

(None reported.)

A.51.4. CLIMATE CHANGE ACTIONS

(None reported.)

A.51.5. RESEARCH ACTIVITIES AND INFORMATION USED

(None reported.)

A.51.6. RESEARCH NEEDS

(None reported.)

APPENDIX B

STATE AQUATIC INVASIVE SPECIES MANAGEMENT PLAN SUMMARIES

CONTENTS

CONTENTS (continued)

CONTENTS (continued)

CONTENTS (continued)

LIST OF TABLES

B.1.   METHODS

We reviewed state aquatic invasive species (AIS) management plans, where available, and assessed how the state addresses climate change specifically, as well as how they generally provide for adaptation of strategies and actions under changing conditions. There are a total of 25 state plans, including 23 AIS-specific plans and 2 general invasive species management plans with a significant AIS focus. As noted in Appendix A, as of 2007, several other states currently are in the process of developing AIS management plans. State plans generally refer to AIS as aquatic nuisance species (ANS). To maintain consistency with state plan language, this appendix generally uses ANS as a synonym for AIS. State plans examined include the following:

- Alaska ANS Management Plan

- Arizona ANS Management Plan

- Connecticut ANS Management Plan

- Hawaii AIS Management Plan

- Idaho Action Plan for Invasive Species

- Illinois State Comprehensive Management Plan

- Indiana ANS Management Plan

- Iowa Plan for the Management of ANS in Iowa

- Kansas ANS Management Plan

- Louisiana State Management Plan for AIS in Louisiana

- Maine Action Plan for Managing Invasive Aquatic Species

- Massachusetts AIS Management Plan

- Michigan ANS State Management Plan

- Missouri ANS Management Plan

- Montana ANS Management Plan

- New York Non-indigenous Aquatic Species Comprehensive Management Plan

- North Dakota ANS Management Plan

- Ohio Comprehensive Management Plan for ANS

- Oregon ANS Management Plan

- Pennsylvania AIS Management Plan

- South Carolina Aquatic Plant Management Plan Part I and II

- Texas State Comprehensive Management Plan for ANS

- Virginia Invasive Species Management Plan

- Washington State ANS Management Plan

- Wisconsin Comprehensive Management Plan to Prevent Further Introductions and Control Existing Populations of AIS

In summaries that follow, we also provide recommendations for revising the various management plans to incorporate climate considerations and management strategies to adapt to climate change; we list the recommendations for individual plan goals and strategies.

B.2. ALASKA AQUATIC NUISANCE SPECIES MANAGEMENT PLAN

B.2.1. GENERAL DESCRIPTION OF ALASKA'S PLAN

Alaska's Aquatic Nuisance Species (ANS) Management Plan was written by the Alaska Department of Fish and Game and released in October 2002 (available at http://www.adfg.state.ak.us/special/invasive/ak_ansmp.pdf). The Management Plan focuses on prevention of new introductions and identification of and response to the highest invasive species threats. The Plan describes six goals to (1) coordinate ANS management within Alaska and collaborate with other programs; (2) prevent new ANS introductions; (3) detect, monitor, contain, reduce or eradicate ANS; (4) educate the public about ANS prevention and impact reduction; (5) identify, develop, conduct, and disseminate research on Alaskan ANS concerns; and (6) ensure that federal and state regulations promote ANS prevention and control. There are Strategic Actions for each goal and a timetable to complete these actions.

B.2.2. CLIMATE CHANGE AND AQUATIC INVASIVE SPECIES IN ALASKA

Climate models project temperature increases in the Arctic of 1.5 to 5°F (1 to 3°C) by 2030, and 5 to 18°F (3 to 10°C) by 2100, with higher magnitudes of warming in the north and in the winter. Precipitation is projected to increase in most of Alaska, up to 20 to 25% in the north and northwest; however, a 10% decrease in precipitation is projected along the south coast (Parson, 2001a). Permafrost thawing is projected to accelerate. Continued loss of sea ice, with year-round ice disappearing completely in one model by 2100, is also projected. Loss of sea ice allows larger storm surges to develop, increasing erosion and coastal inundation (Parson, 2001a).

These climate-change effects may allow species once limited by Alaska's cold climate to establish. Alaska's Management Plan identifies the green crab (*Carcinus maenas*) as one of the state's highest potential invasive threats. Currently, though, the species is thought to be limited from establishment in part due to cold water temperatures. Warming may allow this species' range to expand to Alaska. The melting permafrost may increase nutrient supply into aquatic systems, increasing susceptibility to invasion by species previously limited by lack of nutrient availability, such as the fish pathogen Whirling disease (*Myxobolus cerebralis*), whose vectors require a more nutrient-rich environment than the state's freshwater streams currently provide.

B.2.3. THE ALASKA PLAN'S CURRENT INTEGRATION OF CLIMATE CHANGE

Table B-1 summarizes how the Alaska ANS Management Plan addresses and incorporates the projected effects of climate change. Although Alaska's Management Plan does not specifically address climate change, the Plan includes descriptions of climate zones and changing conditions that can affect ANS ranges.

Table B-1. Assessment of the Alaska Aquatic Nuisance Species Management Plan

Aspects of plan that may incorporate climate change	Score
Understanding and incorporating potential impacts resulting from climate change 0 = no; 1 = briefly mentions; 2 = includes general discussion; 3 = includes quantitative info and/or specific examples	
Plan specifically mentions climate change	0
Plan acknowledges climatic boundaries of species	2
Plan demonstrates understanding of species and/or ecosystem sensitivity to changing conditions	2
Plan identifies research on the potential effects of species responding to changing conditions	0
Plan acknowledges regional differences in expected climate changes	0
Capacity to adapt to changing conditions 0 = no; 1 = implicitly (i.e., includes goals and strategies that can be used to account for changing conditions, but does not specify changing conditions as part of their purpose); 2 = yes, explicitly, in passing; 3 = yes, explicitly, and specifies associated goals and/or action items	
Plan accounts for changing conditions in its leadership and coordination goals and strategies	1
Plan accounts for changing conditions in its prevention goals and strategies	1
Plan accounts for changing conditions in its early detection/rapid response goals and strategies	0
Plan accounts for changing conditions in its control and management goals and strategies	0
Plan accounts for changing conditions in its restoration goals and strategies	0
Plan accounts for changing conditions in its research goals and strategies	1
Plan accounts for changing conditions in its information management goals and strategies	0
Plan accounts for changing conditions in its education and public awareness goals and strategies	1
Monitoring strategies 0 = no; 1 = yes, briefly mentions; 2 = yes, but unclear how information will be used; 3 = yes, and specifies associated goals and/or action items	
Plan includes strategy to monitor for changing conditions	0
Plan includes strategy to utilize monitoring data	2
Plan includes strategy for managing/updating monitoring data	3
Revision 0 = no; 1 = yes, in passing; 2 = yes, and includes qualitative description; 3 = yes, and includes timeline and/or benchmarks for doing so	
Plan includes strategy for updating and incorporating new information	2
Funding 0 = no; 1 = a source is specified for a portion of the required funding; 2 = a source is specified for a portion of the required funding along with strategies for obtaining remaining funding; 3 = a source is specified for 100% of required funding	
Plan identifies dedicated funding source for implementation	1
Total score:	**16**

B.2.4. INCORPORATING CLIMATE CHANGE INFORMATION

In light of the significant effects of climate change predicted for Alaska, the state may consider specifying climate change-related actions and strategies within the ANS Management

Plan. Climate change-related data, criteria, and models could be incorporated into some of the Strategic Actions outlined in the Management Plan.

B.2.4.1.   Leadership and Coordination

Strategic Action 2A1 calls for coordination and development of action plans for current high priority species or pathways. The action item also calls for coordination on preparing risk assessments to determine additional ANS priority threats. Because species' abilities to spread are affected in part by climate, action plans should incorporate projected changes in water and air temperatures in risk assessments. Invasion pathways linked to human activities also may be sensitive to climate change. For example, recreational boating may increase as the climate warms, which will provide additional invasive species transport opportunities. Therefore, action plans addressing pathways should also incorporate information on climate-change effects.

B.2.4.2.   Prevention

Strategic Action 2A3 recommends that Alaska prohibit, control, or permit the importation of non-native aquatic species based on their invasive potential. Criteria used to identify potentially invasive species should account for projected changes in temperature, nutrient availability, hydrology, and other climate change-related ecological impacts that could modify potential habitat for invasive species previously limited by these factors.

B.2.4.3.   Early Detection/Rapid Response, Control, and Management

Strategic Action 3A1, designed to detect, monitor, contain, reduce, or eradicate populations of ANS as quickly as possible, calls for monitoring waters vulnerable to new ANS introductions and tracking existing populations' distributions. Considering climate change in these assessments may help determine which waters are vulnerable to species invasions. Monitoring data from neighboring states may also allow state staff to track invasive species spreading in response to climate change.

B.2.4.4.   Research

Strategic Action 5A1 recommends an assessment of risks posed to human health, ecosystems, and the economy by ANS introductions. The Management Plan recommends characterizing resources and habitats containing ecological communities that are highly sensitive to invasion. Incorporating climate-change effects into these assessments may strengthen their results.

B.3. ARIZONA AQUATIC NUISANCE SPECIES MANAGEMENT PLAN

B.3.1. GENERAL DESCRIPTION OF ARIZONA'S PLAN

Arizona's Aquatic Nuisance Species (ANS) Management Plan was written by the University of Arizona's Agriculture Department and released in May 2002; however, as of December 2006, it has not been finalized or approved (available at http://ag.arizona.edu/azaqua/extension/ANS/ArizonaPlan.htm). The Arizona Invasive Species Council has recommended development of an Invasive Species Management Plan, which will include a chapter on ANS based on the draft ANS Plan described here. The Draft Plan aims to improve coordination between ANS management programs and activities. The Draft Plan has three main goals to: (1) prevent new ANS introductions; (2) limit the spread of established ANS populations; and (3) abate harmful impacts resulting from ANS infestations, and lists specific actions related to these goals. The Draft Plan identifies seven priority ANS (zebra mussel, hydrilla, Brazilian elodea, parrotfeather, purple loosestrife, giant salvinia, and water hyacinth) and proposes management actions specific to these species.

B.3.2. CLIMATE CHANGE AND AQUATIC INVASIVE SPECIES IN ARIZONA

Temperatures are expected to rise as much as 5°F (3°C) in Southwestern United States over the next 30 years, and precipitation is projected to decrease significantly by 2100 (Seager et al., 2007). Water resources are projected to become scarcer as the climate changes and the demand on water supplies will increase as the population grows (Seager et al., 2007). This decrease in water availability could favor more drought-tolerant invasive species such as tamarisk. A change in temperature and precipitation may also change the structure and composition of Arizona's sensitive ecosystems and native species (SRAG, 2000). This change could potentially benefit fast-growing, more tropical species such as water hyacinth.

B.3.3. THE ARIZONA PLAN'S CURRENT INTEGRATION OF CLIMATE CHANGE

Table B-2 summarizes how the Arizona ANS Management Plan addresses and incorporates the projected effects of climate change. Arizona's Draft Plan does not specifically address climate change, but does mention climatic boundaries of species and a few elements in the Plan allow for changing conditions to be considered in the implementation of the Plan.

B.3.4. INCORPORATING CLIMATE CHANGE INFORMATION

Incorporating the projected effects of climate change, such as increased water temperatures and decreased water levels in the summer, on both native species' ability to survive

Table B-2. Assessment of the Arizona Aquatic Nuisance Species Management Plan

Aspects of plan that may incorporate climate change	Score
Understanding and incorporating potential impacts resulting from climate change 0 = no; 1 = briefly mentions; 2 = includes general discussion; 3 = includes quantitative info and/or specific examples	
Plan specifically mentions climate change	0
Plan acknowledges climatic boundaries of species	1
Plan demonstrates understanding of species and/or ecosystem sensitivity to changing conditions	0
Plan identifies research on the potential effects of species responding to changing conditions	0
Plan acknowledges regional differences in expected climate changes	0
Capacity to adapt to changing conditions 0 = no; 1 = implicitly (i.e., includes goals and strategies that can be used to account for changing conditions, but does not specify changing conditions as part of their purpose); 2 = yes, explicitly, in passing; 3 = yes, explicitly, and specifies associated goals and/or action items	
Plan accounts for changing conditions in its leadership and coordination goals and strategies	0
Plan accounts for changing conditions in its prevention goals and strategies	1
Plan accounts for changing conditions in its early detection/rapid response goals and strategies	0
Plan accounts for changing conditions in its control and management goals and strategies	0
Plan accounts for changing conditions in its restoration goals and strategies	0
Plan accounts for changing conditions in its research goals and strategies	1
Plan accounts for changing conditions in its information management goals and strategies	0
Plan accounts for changing conditions in its education and public awareness goals and strategies	0
Monitoring strategies 0 = no; 1 = yes, briefly mentions; 2 = yes, but unclear how information will be used; 3 = yes, and specifies associated goals and/or action items	
Plan includes strategy to monitor for changing conditions	0
Plan includes strategy to utilize monitoring data	3
Plan includes strategy for managing/updating monitoring data	0
Revision 0 = no; 1 = yes, in passing; 2 = yes, and includes qualitative description; 3 = yes, and includes timeline and/or benchmarks for doing so	
Plan includes strategy for updating and incorporating new information	0*
Funding 0 = no; 1 = a source is specified for a portion of the required funding; 2 = a source is specified for a portion of the required funding along with strategies for obtaining remaining funding; 3 = a source is specified for 100% of required funding	
Plan identifies dedicated funding source for implementation	0*
Total score:	**6**

*Arizona's Plan has not been formally approved and will be incorporated into the state's larger ANS plan, which is under development as of December 2006.

and invasive species' ability to establish or expand, would make the Plan more robust. The following Strategic Actions outlined in the Plan could incorporate climate change considerations.

B.3.4.1. Leadership and Coordination

Strategic Action IA calls for coordination on developing state-specific and regional lists of ANS that have the potential to spread to Arizona's waters. The Management Plan also calls for coordination on identifying existing and potential transport pathways. Task IE1 recommends assessing these transport mechanisms and developing preventative action plans to interrupt pathways. Because species' abilities to spread and become established also are affected by climate, species lists should include aquatic invasive species (AIS) that could be influenced by projected climate changes. Transport pathways linked to human activities that could be sensitive to climate change also should be considered and assessed. For example, recreational boating may increase as climate warms (unless precipitation also decreases), which will provide increased transport opportunities for primary species of concern such as zebra mussels.

B.3.4.2. Prevention

Strategic Action ID recommends developing and maintaining a monitoring program for early detection and prevention of AIS in uninfested watersheds. Accounting for projected effects of climate change, such as increased water temperatures and decreased water levels in the summer, on both native species' ability to survive and invasive species' ability to become established, could help state staff more effectively determine which watersheds may be more vulnerable to invasion under a changing climate. Additionally, collecting information from adjacent states may increase state staff's awareness of climate-related invasive species threats.

B.3.4.3. Early Detection/Rapid Response, Control, and Management

See Section 3.4.2.

B.3.4.4. Research

Strategic Action IF calls for collaboration among state and federal agencies and academic institutions to study and evaluate potential management actions to limit spread of AIS. This assessment and evaluation could also examine how management actions could be adapted in the context of a changing climate and the predicted impacts for Arizona.

B.4. CONNECTICUT AQUATIC NUISANCE SPECIES MANAGEMENT PLAN

B.4.1. GENERAL DESCRIPTION OF CONNECTICUT'S PLAN

Connecticut's Aquatic Nuisance Species (ANS) Management Plan was written by the Connecticut ANS Working Group with public input (available at http://www.ctiwr.uconn.edu/ProjANS/SubmittedMaterial2005/Material200601/ANS%20Plan%20Final%20Draft121905.pdf). The ANS Working Group is composed of state staff from the Connecticut Department of Environmental Protection, Sea Grant College Program, and the Connecticut Institute of Water Resources, as well as other state and regional partners. The Management Plan's primary goal is to establish a comprehensive strategy to minimize the negative impacts of ANS to the state's ecology, economy, and public health. Other goals relate to preventive strategies based on monitoring and early-detection efforts. The Plan catalogues and characterizes existing ANS, including their impacts and costs, and discusses the benefits of planned introductions. Research, resource needs, management programs, and funding sources are also described.

B.4.2. CLIMATE CHANGE AND AQUATIC INVASIVE SPECIES IN CONNECTICUT

Projected increases in annual surface temperatures in the Northeastern region of the United States are projected to average 10°F (5.3°C) by 2070. Nearly all model simulations of future precipitation show consistent increases in winter precipitation and no change to a decrease in summer rainfall. By 2100, precipitation is projected to increase an average of 11 to 14% in the winter. Regional sea surface temperatures are projected to increase in accordance with regional air temperatures; these increasing temperatures have the potential to expand the range of warm-water species northward and permit invasive species to spread into these waters, which had previously been previously too cold to allow for invasive species' survival (Hayhoe et al., 2007).

B.4.3. THE CONNECTICUT PLAN'S CURRENT INTEGRATION OF CLIMATE CHANGE

Table B-3 summarizes how the Connecticut ANS Management Plan addresses and incorporates the projected effects of climate change. While Connecticut's Plan does not include a strong focus on climate change, changing temperatures, shifting winds and currents, and the climatic sensitivities of the region are briefly mentioned in relation to specific species or habitats. Many of the research and management tasks stress the importance of carefully monitoring changing conditions.

Table B-3. Assessment of the Connecticut Aquatic Nuisance Species Management Plan

Aspects of plan that may incorporate climate change	Score
Understanding and incorporating potential impacts resulting from climate change 0 = no; 1 = briefly mentions; 2 = includes general discussion; 3 = includes quantitative info and/or specific examples	
Plan specifically mentions climate change	1
Plan acknowledges climatic boundaries of species	1
Plan demonstrates understanding of species and/or ecosystem sensitivity to changing conditions	1
Plan identifies research on the potential effects of species responding to changing conditions	0
Plan acknowledges regional differences in expected climate changes	0
Capacity to adapt to changing conditions 0 = no; 1 = implicitly (i.e., includes goals and strategies that can be used to account for changing conditions, but does not specify changing conditions as part of their purpose); 2 = yes, explicitly, in passing; 3 = yes, explicitly, and specifies associated goals and/or action items	
Plan accounts for changing conditions in its leadership and coordination goals and strategies	1
Plan accounts for changing conditions in its prevention goals and strategies	1
Plan accounts for changing conditions in its early detection/rapid response goals and strategies	0
Plan accounts for changing conditions in its control and management goals and strategies	1
Plan accounts for changing conditions in its restoration goals and strategies	0
Plan accounts for changing conditions in its research goals and strategies	1
Plan accounts for changing conditions in its information management goals and strategies	0
Plan accounts for changing conditions in its education and public awareness goals and strategies	0
Monitoring strategies 0 = no; 1 = yes, briefly mentions; 2 = yes, but unclear how information will be used; 3 = yes, and specifies associated goals and/or action items	
Plan includes strategy to monitor for changing conditions	0
Plan includes strategy to utilize monitoring data	1
Plan includes strategy for managing/updating monitoring data	1
Revision 0 = no; 1 = yes, in passing; 2 = yes, and includes qualitative description; 3 = yes, and includes timeline and/or benchmarks for doing so	
Plan includes strategy for updating and incorporating new information	1
Funding 0 = no; 1 = a source is specified for a portion of the required funding; 2 = a source is specified for a portion of the required funding along with strategies for obtaining remaining funding; 3 = a source is specified for 100% of required funding	
Plan identifies dedicated funding source for implementation	2
Total score:	**12**

B.4.4. INCORPORATING CLIMATE CHANGE INFORMATION

While changing conditions are acknowledged, the full scope of effects resulting from climate change is not explicitly addressed in the Plan's management strategies. For example, the following Strategic Actions outlined in the Plan could incorporate climate change considerations.

B.4.4.1. Leadership and Coordination

The Management Plan's "Coordinate Beyond Connecticut" (5.1.C) section recognizes that jurisdictional boundaries do not necessarily apply when managing ANS. Strategies related to regional coordination provide an excellent opportunity for information sharing about changing conditions that could allow species to move between habitats.

B.4.4.2. Prevention

Objective 3 in Chapter 5 of the Management Plan outlines prevention actions. The Plan recognizes that the most likely aquatic invasive species (AIS) introductions have already occurred in other Northeast states. When identifying likely species to establish, effects of climate change should be considered.

B.4.4.3. Early Detection/Rapid Response, Control, and Management

The keystone of the Connecticut ANS Plan relies on an expanded monitoring strategy that will allow for the early detection of new infestations, as well as monitoring of existing ANS populations. Monitoring strategies should incorporate climate change information in order to detect species that may arrive in the state's habitats as a result of changing conditions. Considering climate change may increase the effectiveness of early detection/rapid response protocols. Furthermore, regional efforts may become more effective as states document AIS populations, allowing species-specific, rapid response protocols to be developed. Control of existing ANS also may be more effective if changes in water temperature and precipitation patterns are considered.

B.4.4.4. Research

Objective 7 in Chapter 5 of Connecticut's ANS Mangement Plan describes a research strategy that may be modified according to changes in ANS populations. The effects of climate change on Long Island Sound and the state's inland waterways and aquatic habitats should be included as a research priority. The Management Plan also requires the on-going designation of priority species using "improved knowledge of ANS distribution and impacts." When designating priorities, state staff should also consider climate-change effects on species distributions.

B.5.   HAWAII AQUATIC INVASIVE SPECIES MANAGEMENT PLAN

B.5.1.   GENERAL DESCRIPTION OF HAWAII'S PLAN

Hawaii's Aquatic Invasive Species (AIS) Management Plan was developed by the Hawaii Invasive Species Council and the Department of Land and Natural Resources, Division of Aquatic Resources and released in 2003 (available at http://www.state.hi.us/dlnr/dar/pubs/ais_mgmt_plan_final.pdf). The goal of the Management Plan is to minimize harmful impacts of AIS through prevention and management of their introduction, expansion, and dispersal. The Plan identifies specific objectives to achieve this goal, including improved coordination, early detection and rapid response, monitoring of existing AIS, increased education and research, and effective laws promoting prevention and control. The Management Plan includes a monitoring and evaluation program and an implementation table that outlines responsible agencies and funding.

B.5.2.   CLIMATE CHANGE AND AQUATIC INVASIVE SPECIES IN HAWAII

The effects of climate change in the tropical Pacific Basin are expected to cause a gradual warming of sea surface and air temperatures. Climate models project possible increases of 2.5 to 5.5°F (1.4 to 3.1°C) between 2080 and 2099 (Christensen et al., 2007). Model results project a 3% increase in precipitation in the Southern Pacific region. Sea levels are expected to rise in the Pacific Ocean, although the magnitude is uncertain (Christensen et al., 2007). Models project a rise between 3.9 to 4.7 inches (10 to 12 cm) in the short term and 11.8 to 15.0 inches (30 to 38 cm) over the long term (PIRAG, 2001). Climate models also show a gradual increase in tropical cyclone frequency for islands in the Central and East-Central Pacific region (PIRAG, 2001).

Hawaii contains 40% of the United States's endangered species. Invasive species are one of the major threats. For example, warming temperatures may allow mosquitoes to survive at higher altitudes, pushing already threatened native forest birds to higher elevations and into smaller ranges. Coral species and coral reef-dependent species also may be impacted by climate change and invasive species (Harvell et al., 1999). Warming water temperatures are projected to cause coral bleaching and increase the occurrence, the severity, and the spread of marine diseases (Harvell et al., 2002; Jones et al., 2004). Diseases could further exacerbate these problems by weakening ecosystems (Jones et al., 2004).

B.5.3.   THE HAWAII PLAN'S CURRENT INTEGRATION OF CLIMATE CHANGE

Table B-4 summarizes how the Hawaii Aquatic Invasive Species Management Plan addresses and incorporates the projected effects of climate change. Although Hawaii's Plan does not explicitly mention climate change, regular updates to address and adapt to changing circumstances are planned, providing an opportunity to include climate change in future drafts.

Table B-4. Assessment of the Hawaii Aquatic Nuisance Species Management Plan

Aspects of plan that may incorporate climate change	Score
Understanding and incorporating potential impacts resulting from climate change 0 = no; 1 = briefly mentions; 2 = includes general discussion; 3 = includes quantitative info and/or specific examples	
Plan specifically mentions climate change	0
Plan acknowledges climatic boundaries of species	2
Plan demonstrates understanding of species and/or ecosystem sensitivity to changing conditions	2
Plan identifies research on the potential effects of species responding to changing conditions	0
Plan acknowledges regional differences in expected climate changes	0
Capacity to adapt to changing conditions 0 = no; 1 = implicitly (i.e., includes goals and strategies that can be used to account for changing conditions, but does not specify changing conditions as part of their purpose); 2 = yes, explicitly, in passing; 3 = yes, explicitly, and specifies associated goals and/or action items	
Plan accounts for changing conditions in its leadership and coordination goals and strategies	0
Plan accounts for changing conditions in its prevention goals and strategies	0
Plan accounts for changing conditions in its early detection/rapid response goals and strategies	1
Plan accounts for changing conditions in its control and management goals and strategies	0
Plan accounts for changing conditions in its restoration goals and strategies	0
Plan accounts for changing conditions in its research goals and strategies	1
Plan accounts for changing conditions in its information management goals and strategies	0
Plan accounts for changing conditions in its education and public awareness goals and strategies	1
Monitoring strategies 0 = no; 1 = yes, briefly mentions; 2 = yes, but unclear how information will be used; 3 = yes, and specifies associated goals and/or action items	
Plan includes strategy to monitor for changing conditions	0
Plan includes strategy to utilize monitoring data	3
Plan includes strategy for managing/updating monitoring data	3
Revision 0 = no; 1 = yes, in passing; 2 = yes, and includes qualitative description; 3 = yes, and includes timeline and/or benchmarks for doing so	
Plan includes strategy for updating and incorporating new information	1
Funding 0 = no; 1 = a source is specified for a portion of the required funding; 2 = a source is specified for a portion of the required funding along with strategies for obtaining remaining funding; 3 = a source is specified for 100% of required funding	
Plan identifies dedicated funding source for implementation	0
Total score:	**14**

B.5.4. INCORPORATING CLIMATE CHANGE INFORMATION

Climate change is projected to have significant ecological effects in Hawaii, and considering these effects may increase the effectiveness of actions and strategies outlined in the Management Plan. Climate change-related data, criteria, and models could be incorporated into some of the Strategic Actions outlined in the Management Plan

B.5.4.1. Leadership and Coordination

Strategies related to coordination provide an excellent opportunity for information sharing about changing conditions that could allow species to move between habitats.

B.5.4.2. Prevention

Strategy 2B8 calls for identifying "ecologically sensitive" marine and inland waters that have few or no AIS and determining and implementing precautionary actions. Areas not previously considered may be vulnerable to AIS due to climate change effects such as warmer waters, droughts or flooding, or sea level rise. For example, salt water intrusion from sea level rise may harm freshwater ecosystems close to the coast and could allow salt tolerant AIS to thrive and out-compete native species. Thus, climate change considerations should be taken into account when identifying "ecologically sensitive" areas.

B.5.4.3. Early Detection/Rapid Response, Control, and Management

Strategies 3A and 3B call for continuing current monitoring to improve understanding of spatial and temporal distributions of existing species and detect new species. Monitoring should be modified to address how climate change may affect AIS rates of spread in order to more accurately predict distributions and prevent their establishment. Proactive monitoring will increase the efficacy of early detection and rapid response. Strategy 4C recommends integrating knowledge on control and management efforts from Hawaii with national and international information on specific species to develop long-term plans for containment and eradication. State staff should consider how climate change effects (e.g., increased runoff, water temperature, or drought, may impact the success of recommended control and eradication methods.

B.5.4.4. Research

Objective 6 recommends research on the economic impacts of AIS. Climate change should be considered in this research. For example, if coral or fish diseases become more prevalent due to warmer waters, decreasing coral reef abundance, Hawaii's tourism industry could suffer. Coral bleaching due to climate change could further exacerbate this impact.

B.6. IDAHO ACTION PLAN FOR INVASIVE SPECIES

B.6.1. GENERAL DESCRIPTION OF IDAHO'S PLAN

As of December 2006, Idaho's Aquatic Nuisance Species (ANS) Management Plan was under development; however, the state's Action Plan for Invasive Species includes extensive information on ANS and is summarized here (available at http://www.agri.state.id.us/Categories/Environment/InvasiveSpeciesCouncil/documents/Idaho%27s%20Invasive%20Species%20Plan.pdf). The Management Plan, written by the Idaho Invasive Species Council for Governor Kempthorne in 2005, identifies gaps in current state management efforts and explains how these gaps may be addressed. The Management Plan contains 22 actions to address 7 main management approaches, each with a long-term goal: (1) Early intervention (2) Containment, control, and restoration; (3) Education and training; (4) Research and technology transfer; (5) Assurance of adequate funding; (6) Creation of an adequate, effective legal structure; and (7) Coordination of existing programs. All proposed actions have associated short-term goals, a measurable objective, and an implementation timeline.

B.6.2. CLIMATE CHANGE AND AQUATIC INVASIVE SPECIES IN IDAHO

Average warming in the Pacific Northwestern region of the United States is projected to reach 3°F (1.7°C) by the 2020s and 5°F (2.8°C) by the 2050s. Annual precipitation projections are less certain; projected precipitation levels range from a small decrease (7% or 2 inches) to a slightly larger increase (13% or 4 inches). Heavier winter rainfall would increase soil saturation, landslides, and winter flooding. The projected precipitation increases are expected to be concentrated in winter, with decreases or smaller increases during summer; for this reason, even the projections that show increases in annual precipitation show decreases in water availability (Parson, 2001b).

As temperatures increase and water supplies decrease, aquatic invasive species (AIS) such as salt cedar may gain an advantage over native, less tolerant species. Idaho may experience an increase in AIS transported by humans as air and water temperatures warm and water-based recreation increases or is extended for longer periods throughout the year.

B.6.3. THE IDAHO PLAN'S CURRENT INTEGRATION OF CLIMATE CHANGE

Table B-5 summarizes how the Idaho Action Plan for Invasive Species addresses and incorporates the predicted effects of climate change. Idaho's Action Plan for Invasive Species does not incorporate climate change impacts on its management actions.

Table B-5. Assessment of the Idaho Aquatic Nuisance Species Management Plan

Aspects of plan that may incorporate climate change	Score
Understanding and incorporating potential impacts resulting from climate change 0 = no; 1 = briefly mentions; 2 = includes general discussion; 3 = includes quantitative info and/or specific examples	
Plan specifically mentions climate change	0
Plan acknowledges climatic boundaries of species	1
Plan demonstrates understanding of species and/or ecosystem sensitivity to changing conditions	0
Plan identifies research on the potential effects of species responding to changing conditions	0
Plan acknowledges regional differences in expected climate changes	0
Capacity to adapt to changing conditions 0 = no; 1 = implicitly (i.e., includes goals and strategies that can be used to account for changing conditions, but does not specify changing conditions as part of their purpose); 2 = yes, explicitly, in passing; 3 = yes, explicitly, and specifies associated goals and/or action items	
Plan accounts for changing conditions in its leadership and coordination goals and strategies	0
Plan accounts for changing conditions in its prevention goals and strategies	0
Plan accounts for changing conditions in its early detection/rapid response goals and strategies	0
Plan accounts for changing conditions in its control and management goals and strategies	0
Plan accounts for changing conditions in its restoration goals and strategies	0
Plan accounts for changing conditions in its research goals and strategies	0
Plan accounts for changing conditions in its information management goals and strategies	0
Plan accounts for changing conditions in its education and public awareness goals and strategies	0
Monitoring strategies 0 = no; 1 = yes, briefly mentions; 2 = yes, but unclear how information will be used; 3 = yes, and specifies associated goals and/or action items	
Plan includes strategy to monitor for changing conditions	0
Plan includes strategy to utilize monitoring data	0
Plan includes strategy for managing/updating monitoring data	0
Revision 0 = no; 1 = yes, in passing; 2 = yes, and includes qualitative description; 3 = yes, and includes timeline and/or benchmarks for doing so	
Plan includes strategy for updating and incorporating new information	0
Funding 0 = no; 1 = a source is specified for a portion of the required funding; 2 = a source is specified for a portion of the required funding along with strategies for obtaining remaining funding; 3 = a source is specified for 100% of required funding	
Plan identifies dedicated funding source for implementation	0
Total score:	1

B.6.4. INCORPORATING CLIMATE CHANGE INFORMATION

Considering the climate change effects predicted for the State of Idaho, it will be important for the state to incorporate climate change into future revisions of its Invasive Species Action Plan, as well as into the AIS Management Plan currently under development. Climate change-related data, criteria, and models could be incorporated into some aspects of the Idaho Action Plan for Invasive Species.

B.6.4.1. Leadership and Coordination

Strategies related to coordination provide an excellent opportunity for information sharing about changing conditions that could allow species to move between habitats.

B.6.4.2. Prevention

Task 1 in the Early Interventions-Prevention, Early Detection and Rapid Response Section, calls for the creation of lists of high-risk invasive species, or those species that have a high probability of being introduced. Adding climate change considerations into the preparation of these lists can help identify high-risk AIS for targeted prevention activities.

B.6.4.3. Early Detection/Rapid Response, Control, and Management

Task 1 in the Early Interventions-Prevention, Early Detection and Rapid Response Section also calls for the creation of a "red list" comprising species that pose the highest threat. Warming waters and/or decreased water levels not only may influence ecosystem vulnerability, allowing certain species invade and to become established, but also may allow for previously limited invasive species' ranges to expand. These possibilities should be considered in developing a list of high-risk species. Additionally, collecting available monitoring data from neighboring states may allow state staff to track invasive species that are spreading as a result of climate change. This information will also be useful in developing a statewide system for early detection and rapid response, as described in Task 2. The system will utilize scientific protocols to determine the risks posed by invasive species.

B.6.4.4. Research

Task 1 under the Broadening Knowledge through Research and Technology Transfer Section, calls for the identification and prioritization of invasive species research. The Plan highlights species risk assessments to identify habitats susceptible to invasion and to assess potential damages. Because some ecosystems may be more susceptible to invasions as a result of climate change, risk assessments would be more accurate if they consider the projected effects of climate change.

B.7.   ILLINOIS STATE COMPREHENSIVE MANAGEMENT PLAN

B.7.1.   GENERAL DESCRIPTION OF ILLINOIS'S PLAN

Illinois's Management Plan was written by the Illinois Department of Natural Resources and Illinois-Indiana Sea Grant in 1999 (available at http://www.anstaskforce.gov/State%20Plans/ilansplan.pdf). The Management Plan has three primary goals to prevent new aquatic invasive species (AIS) into the Great Lakes and Mississippi River Basin, to limit the spread of established AIS, and to abate impacts by currently established AIS. The Plan includes a list of AIS management tasks for state waters. In addition, arguing that the costs of AIS to industry and resource production far outweigh the cost of conducting AIS management, the Management Plan calls for the commitment of state staff and resources to address AIS and implement prescribed management tasks. However, the Management Plan leaves specific budgeting for a future work plan.

B.7.2.   CLIMATE CHANGE AND AQUATIC INVASIVE SPECIES IN ILLINOIS

Temperatures are projected to increase by 5 to10°F (2.8 to 5.5°C) in the Midwest region throughout the 21st century. Precipitation is expected to increase by approximately 10 to 30% across the region. Increasing temperatures are expected to increase evaporation, triggering a soil moisture deficit, reduction in lake and river levels, and more drought-like conditions in much of the region. For smaller lakes and rivers, reduced flows are likely to intensify water quality issues. In particular, eutrophication of lakes will likely increase due to increases in excess nutrient runoff from heavy precipitation events and warmer lake temperatures that stimulate algae growth (Easterling and Karl, 2001).

As water temperatures in lakes increase, significant changes in freshwater ecosystems will occur. For example, a shift from cold-water fish species such as trout, to warmer water species, like bass and catfish could take place. Warmer waters also may create an environment that is more susceptible to invasions by non-native species (Easterling and Karl, 2001).

B.7.3.   THE ILLINOIS PLAN'S CURRENT INTEGRATION OF CLIMATE CHANGE

Table B-6 summarizes how the Illinois's Management Plan addresses and incorporates the projected effects of climate change. Illinois' Management Plan notes that habitat changes are imminent and related to AIS, but it does not discuss climate change as a possible cause of changes. The Plan is designed to be generally adaptable in a changing environment and allows for continuous reassessment of strategies and actions.

Table B-6. Assessment of the Illinois Aquatic Nuisance Species Management Plan

Aspects of plan that may incorporate climate change	Score
Understanding and incorporating potential impacts resulting from climate change 0 = no; 1 = briefly mentions; 2 = includes general discussion; 3 = includes quantitative info and/or specific examples	
Plan specifically mentions climate change	0
Plan acknowledges climatic boundaries of species	1
Plan demonstrates understanding of species and/or ecosystem sensitivity to changing conditions	1
Plan identifies research on the potential effects of species responding to changing conditions	0
Plan acknowledges regional differences in expected climate changes	0
Capacity to adapt to changing conditions 0 = no; 1 = implicitly (i.e., includes goals and strategies that can be used to account for changing conditions, but does not specify changing conditions as part of their purpose); 2 = yes, explicitly, in passing; 3 = yes, explicitly, and specifies associated goals and/or action items	
Plan accounts for changing conditions in its leadership and coordination goals and strategies	0
Plan accounts for changing conditions in its prevention goals and strategies	0
Plan accounts for changing conditions in its early detection/rapid response goals and strategies	0
Plan accounts for changing conditions in its control and management goals and strategies	0
Plan accounts for changing conditions in its restoration goals and strategies	0
Plan accounts for changing conditions in its research goals and strategies	2
Plan accounts for changing conditions in its information management goals and strategies	0
Plan accounts for changing conditions in its education and public awareness goals and strategies	0
Monitoring strategies 0 = no; 1 = yes, briefly mentions; 2 = yes, but unclear how information will be used; 3 = yes, and specifies associated goals and/or action items	
Plan includes strategy to monitor for changing conditions	0
Plan includes strategy to utilize monitoring data	1
Plan includes strategy for managing/updating monitoring data	1
Revision 0 = no; 1 = yes, in passing; 2 = yes, and includes qualitative description; 3 = yes, and includes timeline and/or benchmarks for doing so	
Plan includes strategy for updating and incorporating new information	0
Funding 0 = no; 1 = a source is specified for a portion of the required funding; 2 = a source is specified for a portion of the required funding along with strategies for obtaining remaining funding; 3 = a source is specified for 100% of required funding	
Plan identifies dedicated funding source for implementation	0
Total score:	**6**

B.7.4. INCORPORATING CLIMATE CHANGE INFORMATION

Illinois could strengthen its AIS strategy by incorporating climate change considerations into regional planning goals and the development of a statewide database to record information on habitat. Climate change-related data, criteria, and models could be incorporated into some aspects of the Illinois State Comprehensive Management Plan.

B.7.4.1. Leadership and Coordination

The Nonindigenous Aquatic Nuisance Prevention and Control Act of 1990 calls for a Great Lakes panel to convene and organize on AIS. The coordinated regional effort addresses regional priorities, recommendations to the federal Aquatic Nuisance Species (ANS) Task Force, and additional federal communications and recommendations. However, the Illinois Management Plan does not outline specific responsibilities for the state in this effort. Illinois and the Great Lakes panel could bolster their efforts by communicating on the anticipated effects of climate change, vectors and species potentially invasive as a result of changing conditions, as well as management strategies adapted to account for changing conditions.

B.7.4.2. Prevention

The Illinois Management Plan's prevention strategies, outlined in Strategic Action 2-2, focus on vectors and barriers to physical dispersal. Prevention strategies may need to be adapted in light of fluctuating lake levels, water temperatures, increased vectors, and other changes influenced by climate. These considerations will ensure the robustness of AIS prevention management decisions and efficient use of scarce resources.

B.7.4.3. Early Detection/Rapid Response, Control, and Management

Illinois has not yet developed an ED/RR protocol but has listed it as a requirement in the state AIS Plan. Changing conditions should be considered as the ED/RR system is developed in order to increase the effectiveness of monitoring strategies for high priority species. Furthermore, climate change may influence which species are determined to be of high risk— thus, climate change information should be considered in developing AIS priority lists.

B.7.4.4. Research

The Illinois Management Plan sets more effective control strategies and better dispersal barriers as research priorities. This research should include adapting management strategies to address the anticipated effects of climate change.

B.8. INDIANA AQUATIC NUISANCE SPECIES MANAGEMENT PLAN

B.8.1. GENERAL DESCRIPTION OF INDIANA'S PLAN

The Indiana Aquatic Nuisance Species (ANS) Management Plan was written by the Indiana Department of Natural Resources and posted in 2003 (available at http://www.in.gov/dnr/invasivespecies/inansmanagementplan.html). The Management Plan has three main goals to (1) prevent aquatic invasive species (AIS) introductions, (2) limit the spread of established AIS, and (3) reduce impacts of established AIS. The Plan also outlines strategies for state and local government agencies and concerned community and research organizations to control AIS infestations in a safe and effective manner. No comprehensive survey of the AIS populations in Indiana has been conducted to date, and so the Plan emphasizes building a foundation of information as a priority. Guiding principles for the Management Plan include strong leadership, provision of resources and state staff, illustration of economic and environmental damages due to AIS, and implementation at all levels of government.

B.8.2. CLIMATE CHANGE AND AQUATIC INVASIVE SPECIES IN INDIANA

Temperatures are projected to increase by 5 to 10°F (2.8 to 5.5°C) in the Midwest region throughout the 21st century. Precipitation is expected to increase by approximately 10 to 30% across the region. Increasing temperatures are expected to increase evaporation, triggering a soil moisture deficit, reduction in lake and river levels, and more drought-like conditions in much of the region. For smaller lakes and rivers, reduced flows are likely to intensify water quality issues. In particular, eutrophication of lakes will likely increase due to increases in excess nutrient runoff from heavy precipitation events and warmer lake temperatures that stimulate algae growth (Easterling and Karl, 2001).

As water temperatures in lakes increase, significant changes in freshwater ecosystems will occur. For example, a shift from cold-water fish species such as trout, to warmer water species, like bass and catfish could take place. Warmer waters also may create an environment that is more susceptible to invasions by non-native species (Easterling and Karl, 2001).

B.8.3. THE INDIANA PLAN'S CURRENT INTEGRATION OF CLIMATE CHANGE

Table B-7 summarizes how the Indiana ANS Management Plan addresses and incorporates the predicted effects of climate change. The Indiana Management Plan includes a broad array of preventative actions, but it does not specifically address how climate change may affect AIS or associated management strategies.

Table B-7. Assessment of the Indiana Aquatic Nuisance Species Management Plan

Aspects of plan that may incorporate climate change	Score
Understanding and incorporating potential impacts resulting from climate change 0 = no; 1 = briefly mentions; 2 = includes general discussion; 3 = includes quantitative info and/or specific examples	
Plan specifically mentions climate change	0
Plan acknowledges climatic boundaries of species	3
Plan demonstrates understanding of species and/or ecosystem sensitivity to changing conditions	0
Plan identifies research on the potential effects of species responding to changing conditions	0
Plan acknowledges regional differences in expected climate changes	0
Capacity to adapt to changing conditions 0 = no; 1 = implicitly (i.e., includes goals and strategies that can be used to account for changing conditions, but does not specify changing conditions as part of their purpose); 2 = yes, explicitly, in passing; 3 = yes, explicitly, and specifies associated goals and/or action items	
Plan accounts for changing conditions in its leadership and coordination goals and strategies	0
Plan accounts for changing conditions in its prevention goals and strategies	1
Plan accounts for changing conditions in its early detection/rapid response goals and strategies	1
Plan accounts for changing conditions in its control and management goals and strategies	0
Plan accounts for changing conditions in its restoration goals and strategies	0
Plan accounts for changing conditions in its research goals and strategies	0
Plan accounts for changing conditions in its information management goals and strategies	0
Plan accounts for changing conditions in its education and public awareness goals and strategies	0
Monitoring strategies 0 = no; 1 = yes, briefly mentions; 2 = yes, but unclear how information will be used; 3 = yes, and specifies associated goals and/or action items	
Plan includes strategy to monitor for changing conditions	0
Plan includes strategy to utilize monitoring data	3
Plan includes strategy for managing/updating monitoring data	0
Revision 0 = no; 1 = yes, in passing; 2 = yes, and includes qualitative description; 3 = yes, and includes timeline and/or benchmarks for doing so	
Plan includes strategy for updating and incorporating new information	3
Funding 0 = no; 1 = a source is specified for a portion of the required funding; 2 = a source is specified for a portion of the required funding along with strategies for obtaining remaining funding; 3 = a source is specified for 100% of required funding	
Plan identifies dedicated funding source for implementation	1
Total score:	**12**

B.8.4. INCORPORATING CLIMATE CHANGE INFORMATION

Indiana could increase the effectiveness and robustness of its Plan by considering anticipated changing conditions that may result from climate change in its AIS management strategies. Climate change-related data, criteria, and models could be incorporated into some aspects of the Indiana ANS Management Plan.

B.8.4.1. Leadership and Coordination

The Management Plan calls for the establishment of consistent methods and priority lists among states that trade regularly or that have overlapping watersheds. Climate change should be a consideration in communication on encroaching species and potential vectors and on adapting management practices to accommodate changing conditions.

B.8.4.2. Prevention

Objective II.A. of the Plan discusses priority vectors for AIS. Climate change should be a consideration in determining vectors and risks of AIS invasion. Not only will species ranges shift as a result of changing conditions, but vectors also may increase AIS dispersal indirectly. For example, recreational boating may increase as the climate warms and waterways remain open for longer periods. AIS risk assessments should also consider anticipated effects of climate change on species' life cycles and pathways in order to identify new species threats. Finally, Indiana's Management Plan emphasizes the importance of monitoring not only high-priority species but also geographic areas at risk. Climate change data should be considered in determining both species and locations at high risk of invasion.

B.8.4.3. Early Detection/Rapid Response, Control, and Management

Objective III.A. of the Management Plan outlines the use of monitoring programs to ensure that invasive species are properly detected, verified, and reported. Because changing water levels and temperatures and precipitation patterns affect AIS's habitat ranges, climate change information should be considered in developing monitoring strategies.

B.8.4.4. Research

Objective V.A. of the Management Plan describes the need for research on and development of control methods for priority species. Control methods, particularly those using biocontrol, should incorporate climate change information to ensure continued efficacy as environmental conditions change.

B.9. PLAN FOR THE MANAGEMENT OF AQUATIC NUISANCE SPECIES IN IOWA

B.9.1. GENERAL DESCRIPTION OF IOWA'S PLAN

Iowa's Department of Natural Resources Eurasian Watermilfoil Program and the Iowa Aquatic Nuisance Species (ANS) Special Task Force led the development of the Plan for the Management of ANS in Iowa (available at http://www.anstaskforce.gov/Iowa-ANS-Mangement-Plan.pdf). The Management Plan was written to guide development of management actions for invasive species, as well as funding mechanisms for prevention, control, and abatement activities for state agencies, local governments, and resource users. Iowa's Management Plan outlines three goals to (1) minimize aquatic invasive species (AIS) introductions, (2) limit the spread of established AIS, and (3) eradicate or control existing AIS to minimize impacts. The Management Plan outlines specific objectives with Strategic Actions to accomplish these goals. The Management Plan also focuses on three priority AIS, but notes that, as the state AIS program evolves, it will incorporate more species. The Plan also includes an implementation table and a program monitoring and evaluation table.

B.9.2. CLIMATE CHANGE AND AQUATIC INVASIVE SPECIES IN IOWA

Temperatures are projected to increase by 5 to 14°F (3 to 8°C) in the winter and 9 to 22°F (5 to 12°C) in the summer. These warmer temperatures will lead to earlier spring snowmelt, which, in combination with increased evaporation in the summer months, could lead to a decrease in surface and ground water availability. Although winter and spring precipitation is expected to increase by 30%, summer precipitation is expected to decrease by 10 to 35%— further impacting water supply. Less water could lead to drier soils and droughts in the summer months. Flood control capacity of wetlands and floodplains may also be degraded, which could result in increased flooding in winter and spring months. Additional flooding in these seasons could cause increased sedimentation and pollution into Iowa's waters (Moser et al., 2004).

As water temperatures warm, species currently limited by Iowa's winter temperatures, such as mosquito fish, may begin to invade the state. Other species established further south, such as the spotted gar, may begin to move northward in the Mississippi River (Moser et al., 2004). Eurasian watermilfoil, already a major problem in Iowa, could worsen under climate change conditions as this species is tolerant of a wide range of conditions (Moser et al., 2004).

B.9.3. THE IOWA PLAN'S CURRENT INTEGRATION OF CLIMATE CHANGE

Table B-8 summarizes how the Plan for the Management of ANS in Iowa addresses and incorporates the projected effects of climate change. Although the Plan includes no specific climate change considerations, it does include monitoring and prevention strategies, and climate change considerations could be incorporated into the Strategic Actions outlined in the Plan.

Table B-8. Assessment of the Iowa Aquatic Nuisance Species Management Plan

Aspects of plan that may incorporate climate change	Score
Understanding and incorporating potential impacts resulting from climate change 0 = no; 1 = briefly mentions; 2 = includes general discussion; 3 = includes quantitative info and/or specific examples	
Plan specifically mentions climate change	0
Plan acknowledges climatic boundaries of species	1
Plan demonstrates understanding of species and/or ecosystem sensitivity to changing conditions	0
Plan identifies research on the potential effects of species responding to changing conditions	0
Plan acknowledges regional differences in expected climate changes	0
Capacity to adapt to changing conditions 0 = no; 1 = implicitly (i.e., includes goals and strategies that can be used to account for changing conditions, but does not specify changing conditions as part of their purpose); 2 = yes, explicitly, in passing; 3 = yes, explicitly, and specifies associated goals and/or action items	
Plan accounts for changing conditions in its leadership and coordination goals and strategies	0
Plan accounts for changing conditions in its prevention goals and strategies	0
Plan accounts for changing conditions in its early detection/rapid response goals and strategies	0
Plan accounts for changing conditions in its control and management goals and strategies	0
Plan accounts for changing conditions in its restoration goals and strategies	0
Plan accounts for changing conditions in its research goals and strategies	0
Plan accounts for changing conditions in its information management goals and strategies	0
Plan accounts for changing conditions in its education and public awareness goals and strategies	0
Monitoring strategies 0 = no; 1 = yes, briefly mentions; 2 = yes, but unclear how information will be used; 3 = yes, and specifies associated goals and/or action items	
Plan includes strategy to monitor for changing conditions	0
Plan includes strategy to utilize monitoring data	3
Plan includes strategy for managing/updating monitoring data	0
Revision 0 = no; 1 = yes, in passing; 2 = yes, and includes qualitative description; 3 = yes, and includes timeline and/or benchmarks for doing so	
Plan includes strategy for updating and incorporating new information	2
Funding 0 = no; 1 = a source is specified for a portion of the required funding; 2 = a source is specified for a portion of the required funding along with strategies for obtaining remaining funding; 3 = a source is specified for 100% of required funding	
Plan identifies dedicated funding source for implementation	2
Total score:	**8**

B.9.4. INCORPORATING CLIMATE CHANGE INFORMATION

Projected effects of climate changes can be incorporated into Iowa's Management Plan's strategies and actions—monitoring and prevention strategies in particular. Including these considerations will ensure that the Plan and its associated AIS management actions are effective and efficient in the long term.

B.9.4.1. Leadership and Coordination

Strategic Action 1A3 calls for developing regional partnerships to evaluate regional AIS threats and to coordinate management efforts. Iowa could use this opportunity to work with other states to determine what AIS are moving north as a result of climate change and what more southern states are doing to prevent, control, and manage these AIS. Strategic Action 5A1 also calls for working with partners to share AIS distribution information based on each state's monitoring efforts, which could also help to identify new AIS threats due to climate change.

B.9.4.2. Prevention

Strategic Action 2A2 calls for risk assessments to identify water bodies at high-risk of AIS invasion. An associated task recommends incorporating data on species' life histories and habitat preferences. Water bodies may become more sensitive to specific AIS as water temperatures rise or water levels decrease; thus, it will be important for state staff conducting risk assessments to incorporate these considerations into the analysis. Similarly, Strategic Action 5A2 and associated Task 5A2b call for designing a monitoring program to help limit the spread of AIS. Understanding invasive threats in the context of climate change will ensure that monitoring efforts target appropriate AIS and habitats.

B.9.4.3. Early Detection/Rapid Response, Control, and Management

Strategic Action 6A3 calls for supporting research that identifies effective management actions for successful AIS control and eradication methods in Iowa. Research must examine how management actions should be adapted in the context of a changing climate and the predicted impacts for the state. Task 6A3a also recommends identifying important data needed to control and/or eradicate AIS in Iowa. Climate change data should be integrated as a part of this research because climate change can impact the success of control and eradication methods.

B.9.4.4. Research

See Section 9.4.3.

B.10. KANSAS AQUATIC NUISANCE SPECIES MANAGEMENT PLAN

B.10.1. GENERAL DESCRIPTION OF KANSAS'S PLAN

Kansas's Aquatic Nuisance Species (ANS) Management Plan was developed by the state's ANS committee with the Kansas Department of Wildlife and Parks taking the lead and approved in 2005 (available at http://www.kdwp.state.ks.us/news/fishing/aquatic_nuisance_species/ks_nuisance_species_plan). The purpose of the Management Plan is to guide state agencies, local governments, public and private organizations, and aquatic resource user groups in developing management strategies, designing public awareness and educational materials, and prioritizing aquatic invasive species (AIS) activities. The goals of the Plan are to prevent new introductions of AIS; to prevent the dispersal of established AIS; to minimize effects of AIS; to educate aquatic users about AIS risks; and to support research on AIS and develop systems to disseminate information. The Management Plan outlines several management objectives to achieve these goals and includes a discussion of existing problems with AIS; a summary of federal, regional, and state policies on AIS; a list of non-indigenous species in Kansas; identification of priority AIS; and a discussion of regional AIS threats.

B.10.2. CLIMATE CHANGE AND AQUATIC INVASIVE SPECIES IN KANSAS

Climate models predict that average temperatures in Kansas increase by as much 3°F (1.5°C) in the summer and 4°F (2°C) in the winter by 2030 (Covich et al., 1997). The increase in summer temperatures and increased evaporation may lead to lower stream flows and lake levels. Decreased water levels could affect biodiversity and lower flows and higher temperatures in the summer could concentrate pollutant levels, further impacting aquatic habitats (Covich et al., 1997).

Climate change may negatively impact native species and allow invasive species' ranges to expand across the Great Plains region (Joyce et al., 2001). For example, purple loosestrife, a priority species in Kansas's Management Plan, can withstand shallow flooding, which may provide a competitive advantage over some native aquatic plants as flooding increases with climate change (NPWRC, 2006). Zebra mussels, another priority species, may also benefit from warmer temperatures. Zebra mussels begin spawning when water temperatures warm to about 54°F (12°C) and continue spawning until water temperature drops below this threshold (KDHE, 2007). Increasing temperatures may provide a longer spawning season for this species.

B.10.3. THE KANSAS PLAN'S CURRENT INTEGRATION OF CLIMATE CHANGE

Table B-9 summarizes how the Kansas ANS Management Plan addresses and incorporates the projected effects of climate change. Although Kansas's Management Plan does not address climate change, it does stress the importance of researching the relationship between changing conditions and AIS invasion, establishment, and impacts.

Table B-9. Assessment of the Kansas Aquatic Nuisance Species Management Plan

Aspects of plan that may incorporate climate change	Score
Understanding and incorporating potential impacts resulting from climate change 0 = no; 1 = briefly mentions; 2 = includes general discussion; 3 = includes quantitative info and/or specific examples	
Plan specifically mentions climate change	0
Plan acknowledges climatic boundaries of species	0
Plan demonstrates understanding of species and/or ecosystem sensitivity to changing conditions	0
Plan identifies research on the potential effects of species responding to changing conditions	0
Plan acknowledges regional differences in expected climate changes	0
Capacity to adapt to changing conditions 0 = no; 1 = implicitly (i.e., includes goals and strategies that can be used to account for changing conditions, but does not specify changing conditions as part of their purpose); 2 = yes, explicitly, in passing; 3 = yes, explicitly, and specifies associated goals and/or action items	
Plan accounts for changing conditions in its leadership and coordination goals and strategies	0
Plan accounts for changing conditions in its prevention goals and strategies	0
Plan accounts for changing conditions in its early detection/rapid response goals and strategies	0
Plan accounts for changing conditions in its control and management goals and strategies	0
Plan accounts for changing conditions in its restoration goals and strategies	0
Plan accounts for changing conditions in its research goals and strategies	3
Plan accounts for changing conditions in its information management goals and strategies	0
Plan accounts for changing conditions in its education and public awareness goals and strategies	0
Monitoring strategies 0 = no; 1 = yes, briefly mentions; 2 = yes, but unclear how information will be used; 3 = yes, and specifies associated goals and/or action items	
Plan includes strategy to monitor for changing conditions	0
Plan includes strategy to utilize monitoring data	3
Plan includes strategy for managing/updating monitoring data	3
Revision 0 = no; 1 = yes, in passing; 2 = yes, and includes qualitative description; 3 = yes, and includes timeline and/or benchmarks for doing so	
Plan includes strategy for updating and incorporating new information	3
Funding 0 = no; 1 = a source is specified for a portion of the required funding; 2 = a source is specified for a portion of the required funding along with strategies for obtaining remaining funding; 3 = a source is specified for 100% of required funding	
Plan identifies dedicated funding source for implementation	2
Total score:	**14**

B.10.4. INCORPORATING CLIMATE CHANGE INFORMATION

Given the predicted effects of climate change in the state, staff should consider climate change in its AIS management strategies. Climate change-related data, criteria, and models could be incorporated into some aspects of the Kansas ANS Management Plan.

B.10.4.1. Leadership and Coordination

The Management Plan focuses on prevention and on priority species and identifies collaboration and coordination as Objective 1. Climate change should be a consideration in communication on encroaching species and potential vectors and on adapting management practices to accommodate changing conditions.

B.10.4.2. Prevention

Strategic Action 2A1 calls for the identification of AIS that have the greatest potential to establish in Kansas and the identification of existing and potential pathways that facilitate new AIS introductions. Including climate change considerations into assessments of invasive potential could improve predictions of establishment and range expansions. Strategic Action 2A2 call for the establishment of approaches to facilitate legislative, regulatory, and other actions needed to prevent new AIS introductions and to promote rules that establish the state's authority to control these introductions. The consideration of climate change effects could improve these prevention measures by helping identify species with high invasive potential

B.10.4.3. Early Detection/Rapid Response, Control, and Management

Strategic Action 4A1 includes tasks on researching and developing control strategies using the best available science and coordinating with other entities involved in AIS control. Climate change data should be integrated as a part of this research because climate change can impact the success of control and eradication methods, particularly the consideration of biocontrol organisms.

B.10.4.4. Research

Strategic Action 6A1 requires Kansas to support research that identifies, predicts, and prioritizes potential AIS introductions. Potential introductions are subject to the impacts of climate change. As temperatures, precipitation regimes, and nutrient availability change, previously limited AIS may be allowed to establish in the state. Kansas could support research to understand how changing conditions may influence AIS spread to guide prevention efforts.

B.11. STATE MANAGEMENT PLAN FOR AQUATIC INVASIVE SPECIES IN LOUISIANA

B.11.1. GENERAL DESCRIPTION OF LOUISIANA'S PLAN

The State Management Plan for aquatic invasive species (AIS) in Louisiana was developed by the Louisiana AIS Task Force, which is led by the Louisiana Department of Wildlife and Fisheries (available at http://is.cbr.tulane.edu/docs_IS/Louisiana-AIS-Mgt-Plan.pdf). The Plan describes the nature and extent of the AIS problem in Louisiana and proposes actions to minimize the negative impacts of AIS. The Management Plan's goal is to prevent and control the introduction of new non-native species into Louisiana, to control the spread and impact of existing AIS, and to eradicate established AIS wherever possible. The Management Plan outlines objectives to meet these goals and describes species of concern, pathways of introduction, and existing authorities related to AIS. The Management Plan outlines Strategic Actions and a monitoring and evaluation Plan.

B.11.2. CLIMATE CHANGE AND AQUATIC INVASIVE SPECIES IN LOUISIANA

Climate model projections for the Southeastern region of the United States project a 5.5°F (3°C) increase in annual summer air temperatures and a 6.5°F (3.5°C) increase in winter air temperatures. Mississippi River discharge is projected to increase with climate change, which would most likely increase nutrient loads and water column stratification in the northern Gulf of Mexico, exacerbating the problems of eutrophication and hypoxia (Mulholland et al., 1997). Projections of precipitation changes are conflicting, with one major model predicting a 20% decrease in rainfall versus a 20% increase (LaCoast, 2003). Though sea levels are expected to rise, the precise rate of increase is uncertain. Sea level rise could result in significant coastal wetland loss, increasing open water areas and estuarine depths (Mulholland et al., 1997). Wetland loss will reduce habitat for migratory birds, crayfish, sport fish, and other species.

With rising temperatures and a potential decrease in precipitation, evaporation may also increase, which could result in decreased stream and lake water levels. For example, Louisiana's Management Plan notes that the zebra mussel is not as widespread in the lower Mississippi as it is elsewhere in the country, in part due to the increased stream velocity in the spring that prevents many zebra mussel veligers from attaching to hard substrates. Lower stream flows may allow zebra mussels to become attached in the river and to establish larger communities.

B.11.3. THE LOUISIANA PLAN'S CURRENT INTEGRATION OF CLIMATE CHANGE

Table B-10 summarizes how the State Management Plan for AIS in Louisiana addresses and incorporates the projected effects of climate change. Although Louisiana's Plan does not specifically incorporate climate change, it does recognize that the state's wet, subtropical climate, and long growing season, make it particularly sensitive to AIS invasion.

Table B-10. Assessment of the Louisiana Aquatic Nuisance Species Management Plan

Aspects of plan that may incorporate climate change	Score
Understanding and incorporating potential impacts resulting from climate change 0 = no; 1 = briefly mentions; 2 = includes general discussion; 3 = includes quantitative info and/or specific examples	
Plan specifically mentions climate change	0
Plan acknowledges climatic boundaries of species	3
Plan demonstrates understanding of species and/or ecosystem sensitivity to changing conditions	3
Plan identifies research on the potential effects of species responding to changing conditions	0
Plan acknowledges regional differences in expected climate changes	0
Capacity to adapt to changing conditions 0 = no; 1 = implicitly (i.e., includes goals and strategies that can be used to account for changing conditions, but does not specify changing conditions as part of their purpose); 2 = yes, explicitly, in passing; 3 = yes, explicitly, and specifies associated goals and/or action items	
Plan accounts for changing conditions in its leadership and coordination goals and strategies	0
Plan accounts for changing conditions in its prevention goals and strategies	0
Plan accounts for changing conditions in its early detection/rapid response goals and strategies	1
Plan accounts for changing conditions in its control and management goals and strategies	0
Plan accounts for changing conditions in its restoration goals and strategies	0
Plan accounts for changing conditions in its research goals and strategies	0
Plan accounts for changing conditions in its information management goals and strategies	0
Plan accounts for changing conditions in its education and public awareness goals and strategies	0
Monitoring strategies 0 = no; 1 = yes, briefly mentions; 2 = yes, but unclear how information will be used; 3 = yes, and specifies associated goals and/or action items	
Plan includes strategy to monitor for changing conditions	0
Plan includes strategy to utilize monitoring data	3
Plan includes strategy for managing/updating monitoring data	0
Revision 0 = no; 1 = yes, in passing; 2 = yes, and includes qualitative description; 3 = yes, and includes timeline and/or benchmarks for doing so	
Plan includes strategy for updating and incorporating new information	0
Funding 0 = no; 1 = a source is specified for a portion of the required funding; 2 = a source is specified for a portion of the required funding along with strategies for obtaining remaining funding; 3 = a source is specified for 100% of required funding	
Plan identifies dedicated funding source for implementation	2
Total score:	**12**

B.11.4. INCORPORATING CLIMATE CHANGE INFORMATION

Given the effects of climate change projected for Louisiana, staff should consider climate change in its AIS management strategies. Climate change-related data, criteria, and models could be incorporated into some aspects of the State Management Plan for AIS in Louisiana.

B.11.4.1. Leadership and Coordination

State staff could incorporate climate considerations into the Management Plan's first objective, coordination of AIS management activities or programs and collaboration with other programs. Coordination with regional, national, and international efforts could involve monitoring how climate change alters the range and spread of AIS to better guide prevention efforts and communicating with neighboring regions to identify encroaching species ranges.

B.11.4.2. Prevention

Action 11 under Objective 2, which pertains to education on AIS and pathways, calls for support for the ongoing development of the Vulnerability Index for Invasive Species in Southeastern Louisiana, a geographic information system (GIS)-based index and visualization of vectors. The Vulnerability Index currently is an educational tool but could help state staff identify sensitive areas susceptible to invasion. This Vulnerability Index could incorporate projected changes affecting habitat, such as temperature and precipitation fluctuations, to improve predictions.

B.11.4.3. Early Detection/Rapid Response, Control, and Management

Action 3 under Objective 3, which relates to early detection/rapid response, calls for the development of a "Big River" monitoring program to detect and assess AIS introductions, their movement within the state, and potential for establishment. Assessing species' potential for establishment should account for projected changes in habitat that may occur as a result of climate change. Action 9 under Objective 3 proposes the development of a GIS database of invasive species ranges, habitats, and other relevant geographical data. The incorporation of climate change data into this database would inform any modeling based on these data.

B.11.4.4. Research

Although research is not a specific objective in the Management Plan, research activities are incorporated throughout, particularly in the development of GIS-based tools, monitoring programs, and modeling of range expansions. Incorporating climate change considerations into these activities could improve their long-term efficacy and predictions.

B.12. MAINE ACTION PLAN FOR MANAGING INVASIVE AQUATIC SPECIES

B.12.1. GENERAL DESCRIPTION OF MAINE'S PLAN

Maine's Aquatic Invasive Species (AIS) Management Plan was written by Dominie Consulting and the technical subcommittee of the Interagency Task Force on Invasive Aquatic Plans and Nuisance Species and published in October 2002 (available at http://www.maine.gov/dep/blwq/topic/invasives/invplan02.pdf). The Management Plan emphasizes prevention, followed by control and mitigation for AIS that cannot be prevented. The Management Plan also includes education measures to improve AIS prevention and management, as well as an Advisory List of AIS that identifies pathways and threats. Maine's Plan places identified AIS in management categories that prioritize actions for species, including: prevention and eradication, selective control and/or impact management, no action, and dispute resolution. Specific management tasks and implementation timetables are included in the Management Plan.

B.12.2. CLIMATE CHANGE AND AQUATIC INVASIVE SPECIES IN MAINE

Increases in annual air temperatures in the Northeastern region of the United States are projected to average 9.5°F (5.3°C) by 2070. Nearly all model simulations of future precipitation show increases in winter precipitation (11 to 14% by 2100) and no change to a decrease in summer rainfall. Regional sea surface temperatures are anticipated to increase with regional air temperatures; increasing temperatures have the potential to expand the range of warm water species northward and permit AIS expansion into previously colder waters (Hayhoe et al., 2007).

Maine's freshwater systems are relatively remote and currently face less of a threat from AIS than do its marine ecosystems, which are linked to the Great Lakes via the Saint Lawrence Seaway and experience Gulf Stream currents. As waters warm and ocean currents change with climate change, Maine's fresh and marine waters could become more vulnerable to AIS. For example, species that cannot overwinter may become able to survive with warmer waters. In addition, species that previously never entered Maine's waters could now arrive in the state as conditions change. Recreational boating also may increase with warmer temperatures thereby introducing additional species not found in the state at this time, such as the water chestnut.

B.12.3. THE MAINE PLAN'S CURRENT INTEGRATION OF CLIMATE CHANGE

Table B-11 summarizes how the Maine Action Plan for Managing Invasive Aquatic Species incorporates the projected effects of climate change. Maine's Plan acknowledges that the spread of AIS can be accelerated by climate change; however, the Plan's prevention, control, and monitoring strategies do not reflect this consideration.

Table B-11. Assessment of the Maine Aquatic Nuisance Species Management Plan

Aspects of plan that may incorporate climate change	Score
Understanding and incorporating potential impacts resulting from climate change 0 = no; 1 = briefly mentions; 2 = includes general discussion; 3 = includes quantitative info and/or specific examples	
Plan specifically mentions climate change	1
Plan acknowledges climatic boundaries of species	2
Plan demonstrates understanding of species and/or ecosystem sensitivity to changing conditions	2
Plan identifies research on the potential effects of species responding to changing conditions	0
Plan acknowledges regional differences in expected climate changes	0
Capacity to adapt to changing conditions 0 = no; 1 = implicitly (i.e., includes goals and strategies that can be used to account for changing conditions, but does not specify changing conditions as part of their purpose); 2 = yes, explicitly, in passing; 3 = yes, explicitly, and specifies associated goals and/or action items	
Plan accounts for changing conditions in its leadership and coordination goals and strategies	0
Plan accounts for changing conditions in its prevention goals and strategies	0
Plan accounts for changing conditions in its early detection/rapid response goals and strategies	0
Plan accounts for changing conditions in its control and management goals and strategies	0
Plan accounts for changing conditions in its restoration goals and strategies	0
Plan accounts for changing conditions in its research goals and strategies	0
Plan accounts for changing conditions in its information management goals and strategies	0
Plan accounts for changing conditions in its education and public awareness goals and strategies	0
Monitoring strategies 0 = no; 1 = yes, briefly mentions; 2 = yes, but unclear how information will be used; 3 = yes, and specifies associated goals and/or action items	
Plan includes strategy to monitor for changing conditions	1
Plan includes strategy to utilize monitoring data	0
Plan includes strategy for managing/updating monitoring data	0
Revision 0 = no; 1 = yes, in passing; 2 = yes, and includes qualitative description; 3 = yes, and includes timeline and/or benchmarks for doing so	
Plan includes strategy for updating and incorporating new information	3
Funding 0 = no; 1 = a source is specified for a portion of the required funding; 2 = a source is specified for a portion of the required funding along with strategies for obtaining remaining funding; 3 = a source is specified for 100% of required funding	
Plan identifies dedicated funding source for implementation	0
Total score:	**9**

B.12.4. INCORPORATING CLIMATE CHANGE INFORMATION

Given the projected effects of climate change, especially warmer waters and sea level rise, the Management Plan should include these effects in its strategies and tasks. Climate change-related data, criteria, and models could be incorporated into some aspects of the Action Plan for Managing Invasive Aquatic Species in Maine.

B.12.4.1. Leadership and Coordination

The Management Plan emphasizes prevention and priority species and identifies leadership, coordination, and program monitoring as Objective 1. Climate change should be a consideration in communication on encroaching species and potential vectors and on adapting management practices to accommodate changing conditions. Strategy 2C recommends informing key groups on how to prevent AIS introductions and spread. The Management Plan specifically notes that people associated with water craft transport or releasing species should be targeted. Species transported by recreational or commercial boats or released through other vectors may become increasingly able to survive in Maine as conditions change. Thus, it is important to include this information in education and outreach materials and coordinate with these groups.

B.12.4.2. Prevention

The foundation of the Management Plan's AIS prevention strategy is to determine which species and pathways present the largest threat to Maine's aquatic resources. Although the Management Plan recognized that climate change may impact pathways, the Plan does not incorporate specific climate change considerations into its monitoring and prevention strategies. Incorporating these considerations into AIS and pathway assessments may improve efficacy.

B.12.4.3. Early Detection/Rapid Response, Control, and Management

Strategy 4B1 calls for a more comprehensive and detailed approach to rapid response. Considering how climate change may influence early detection and rapid response efforts could influence the development of this approach. For example, a lake in Maine's northern forests that was previously too cold for water chestnut to overwinter may warm enough to allow this AIS to establish. Considering climate change effects could improve survey methods for early detection.

B.12.4.4. Research

Strategy 5B1 calls for anticipating AIS impacts and researching and developing tools to address them. Research on how climate change will affect established and potential AIS is critical to successful prevention, control, and eradication efforts.

B.13. MASSACHUSETTS AQUATIC INVASIVE SPECIES MANAGEMENT PLAN

B.13.1. GENERAL DESCRIPTION OF MASSACHUSETTS'S PLAN

The Massachusetts Aquatic Invasive Species (AIS) Management Plan was created by the Massachusetts AIS Working Group and released in 2002 (available at http://www.mass.gov/czm/invasives/docs/invasive_species_plan.pdf). The overarching goal is to minimize impacts of AIS in both marine and freshwater environments. The Plan relies on eight objectives to accomplish these goals, including coordination, prevention of new introductions, monitoring new AIS and spread of established AIS, detection and eradication, control, educations, research, and identification of legislative needs. Massachusetts's Management Plan outlines ninety-nine specific tasks and includes a budget to ensure these tasks are implemented.

B.13.2. CLIMATE CHANGE AND AQUATIC INVASIVE SPECIES IN MASSACHUSETTS

Increases in annual air temperatures in the Northeastern region of the United States are projected to average 9.5°F (5.3°C) by 2070. Nearly all model simulations of future precipitation show increases in winter precipitation (11 to 14% by 2100) and no change to a decrease in summer rainfall. Regional sea surface temperatures are anticipated to increase with regional air temperatures; increasing temperatures have the potential to expand the range of warm water species northward and permit AIS expansion into previously colder waters (Hayhoe et al., 2007).

State staff are concerned about the potential establishment of various aquatic plants, including parrot feather, European frog-bit, and giant salvinia. These and other AIS are moving up the East Coast; however, they cannot currently survive Massachusetts's winter temperatures. As waters warm, these species may be able to overwinter and establish in the state.

B.13.3. THE MASSACHUSETTS PLAN'S CURRENT INTEGRATION OF CLIMATE CHANGE

Table B-12 summarizes how the Massachusetts AIS Management Plan addresses and incorporates the projected effects of climate change. Although the Massachusetts Management Plan does not explicitly mention climate change, it does address the importance of monitoring for changing conditions.

B.13.4. INCORPORATING CLIMATE CHANGE INFORMATION

Massachusetts will experience various impacts from climate change and should incorporate climate change concerns into its strategies and actions. Climate change-related data, criteria, and models could be incorporated into some aspects of the Massachusetts AIS Management Plan.

Table B-12. Assessment of the Massachusetts Aquatic Nuisance Species Management Plan

Aspects of plan that may incorporate climate change	Score
Understanding and incorporating potential impacts resulting from climate change 0 = no; 1 = briefly mentions; 2 = includes general discussion; 3 = includes quantitative info and/or specific examples	
Plan specifically mentions climate change	1
Plan acknowledges climatic boundaries of species	3
Plan demonstrates understanding of species and/or ecosystem sensitivity to changing conditions	1
Plan identifies research on the potential effects of species responding to changing conditions	0
Plan acknowledges regional differences in expected climate changes	0
Capacity to adapt to changing conditions 0 = no; 1 = implicitly (i.e., includes goals an1d strategies that can be used to account for changing conditions, but does not specify changing conditions as part of their p1urpose); 2 = yes, explicitly, in passing; 3 = yes, explicitly, and specifies associated goals and/or action items0	
Plan accounts for changing conditions in its leadership and coordination goals and strategies	0
Plan accounts for changing conditions in its prevention goals and strategies	0
Plan accounts for changing conditions in its early detection/rapid response goals and strategies	0
Plan accounts for changing conditions in its control and management goals and strategies	0
Plan accounts for changing conditions in its restoration goals and strategies	0
Plan accounts for changing conditions in its research goals and strategies	0
Plan accounts for changing conditions in its information management goals and strategies	0
Plan accounts for changing conditions in its education and public awareness goals and strategies	0
Monitoring strategies 0 = no; 1 = yes, briefly mentions; 2 = yes, but unclear how information will be used; 3 = yes, and specifies associated goals and/or action items	
Plan includes strategy to monitor for changing conditions	0
Plan includes strategy to utilize monitoring data	3
Plan includes strategy for managing/updating monitoring data	0
Revision 0 = no; 1 = yes, in passing; 2 = yes, and includes qualitative description; 3 = yes, and includes timeline and/or benchmarks for doing so	
Plan includes strategy for updating and incorporating new information	0
Funding 0 = no; 1 = a source is specified for a portion of the required funding; 2 = a source is specified for a portion of the required funding along with strategies for obtaining remaining funding; 3 = a source is specified for 100% of required funding	
Plan identifies dedicated funding source for implementation	2
Total score:	**10**

B.13.4.1. Leadership and Coordination

The Management Plan identifies coordination as Objective 1. Climate change should be a consideration in communication on encroaching species and potential vectors and on adapting management practices to accommodate changing conditions. Task 6C calls for developing outreach materials with information on transport vectors. Information about how climate change will affect these vectors and specific AIS should be included so that people associated with these vectors better understand the scope of the problem.

B.13.4.2. Prevention

Objective 2 identifies prevention tasks associated with specific vectors. Information on how climate change may affect these vectors could improve the efficacy of prevention programs.

B.13.4.3. Early Detection/Rapid Response, Control, and Management

Task 5B1 calls for an increase in research on effective biocontrol methods for AIS in Massachusetts's waters. Consideration of projected climate change impacts is necessary to ensure these controls will remain effective. In addition, when developing a management priorities list (Task 5C), state staff should consider how climate change may affect the spread and establishment of AIS to best target these management efforts. Task 5D4 calls for reintroducing native species as part of a restoration program for lakes and ponds. Given that climate change can alter habitats, it will be important to consider how these effects will influence native species and habitats and AIS. A restoration plan should focus on native species that can thrive under or withstand climate change. These considerations may make habitats more robust and less vulnerable to potential invasions as conditions change. Integrating this information into a restoration plan/program also will make restoration activities more successful.

B.13.4.4. Research

Task 7A1 recommends that the Massachusetts's leading scientists and state staff determine the state's research priorities. This task provides an opportunity for these scientists and state staff to include research on climate change effects on AIS as a part of the agenda.

B.14. MICHIGAN AQUATIC NUISANCE SPECIES STATE MANAGEMENT PLAN: UPDATE

B.14.1. GENERAL DESCRIPTION OF MICHIGAN'S PLAN

The Michigan Aquatic Nuisance Species (ANS) Management Plan was prepared by the Michigan Office of Great Lakes and the Department of Environmental Quality and was published in 2002 (available at http://www.deq.state.mi.us/documents/deq-ogl-ANSPlan2002.pdf). The Management Plan updates the state's 1996 Non-indigenous ANS State Management Plan. The Plan recommends three implementation actions to (1) coordinate policies and legislation to reduce ANS impacts, (2) develop materials and activities addressing prevention, control, monitoring, research and policy making, and (3) establish a collaborative network for ANS research and monitoring. The Management Plan states that more work is needed to prevent and control ANS, but notes the progress made since the 1996 Plan. Many vectors could bring ANS into Michigan, and the Management Plan includes specific strategies to better address this issue.

B.14.2. CLIMATE CHANGE AND AQUATIC INVASIVE SPECIES IN MICHIGAN

Climate change models project that temperature increases over the next century could result in an earlier peak snow melt in the spring and higher evaporation rates in the summer months. These factors could lead to lower inland stream flow levels in the summer and reduced water flow into the upper Great Lakes. Lower water levels may impede shipping traffic and prevent more introductions through ballast water releases; however, changes in water flows also could alter native habitats making them more vulnerable to ANS (Magnuson et al., 1997).

Certain ANS may shift their ranges in response to climate change. State staff are concerned about species such as the Asian carp finding their way into state waters. An electrical barrier in the canal separating the Mississippi River from Lake Michigan currently prevents the Asian carp from entering Michigan waters. If water levels are altered by climate change, the efficacy of the barrier could be reduced, allowing the Asian carp to pass into state waters.

B.14.3. THE MICHIGAN PLAN'S CURRENT INTEGRATION OF CLIMATE CHANGE

Table B-13 summarizes how the Michigan Plan addresses and incorporates the projected effects of climate change. The Michigan Management Plan recognizes the need to closely monitor state waters and carefully coordinate control activities and information sharing. The Plan update also refers to focusing prevention and management strategies on the most likely invaders. Although the Management Plan does not include climate change considerations in its strategies, its framework would allow for incorporation of this information.

Table B-13. Assessment of the Michigan Aquatic Nuisance Species Management Plan

Aspects of plan that may incorporate climate change	Score
Understanding and incorporating potential impacts resulting from climate change 0 = no; 1 = briefly mentions; 2 = includes general discussion; 3 = includes quantitative info and/or specific examples	
Plan specifically mentions climate change	0
Plan acknowledges climatic boundaries of species	1
Plan demonstrates understanding of species and/or ecosystem sensitivity to changing conditions	0
Plan identifies research on the potential effects of species responding to changing conditions	0
Plan acknowledges regional differences in expected climate changes	0
Capacity to adapt to changing conditions 0 = no; 1 = implicitly (i.e., includes goals and strategies that can be used to account for changing conditions, but does not specify changing conditions as part of their purpose); 2 = yes, explicitly, in passing; 3 = yes, explicitly, and specifies associated goals and/or action items	
Plan accounts for changing conditions in its leadership and coordination goals and strategies	0
Plan accounts for changing conditions in its prevention goals and strategies	0
Plan accounts for changing conditions in its early detection/rapid response goals and strategies	0
Plan accounts for changing conditions in its control and management goals and strategies	0
Plan accounts for changing conditions in its restoration goals and strategies	0
Plan accounts for changing conditions in its research goals and strategies	0
Plan accounts for changing conditions in its information management goals and strategies	0
Plan accounts for changing conditions in its education and public awareness goals and strategies	0
Monitoring strategies 0 = no; 1 = yes, briefly mentions; 2 = yes, but unclear how information will be used; 3 = yes, and specifies associated goals and/or action items	
Plan includes strategy to monitor for changing conditions	0
Plan includes strategy to utilize monitoring data	3
Plan includes strategy for managing/updating monitoring data	0
Revision 0 = no; 1 = yes, in passing; 2 = yes, and includes qualitative description; 3 = yes, and includes timeline and/or benchmarks for doing so	
Plan includes strategy for updating and incorporating new information	0
Funding 0 = no; 1 = a source is specified for a portion of the required funding; 2 = a source is specified for a portion of the required funding along with strategies for obtaining remaining funding; 3 = a source is specified for 100% of required funding	
Plan identifies dedicated funding source for implementation	0
Total score:	**4**

B.14.4. INCORPORATING CLIMATE CHANGE INFORMATION

Climate change will affect aquatic ecosystems in Michigan and any ANS Plans should incorporate climate change into strategies and actions. Climate change-related data, criteria, and models could be incorporated into some aspects of the Michigan ANS Management Plan.

B.14.4.1. Leadership and Coordination

Three of Michigan's borders are on the upper Great Lakes, and the state historically has taken the lead on invasive species issues such as ballast water regulation. Coordination with other states in the region is a cornerstone of the Management Plan. Goal III on information and education calls on Great Lakes regional policymakers and user groups to promote ANS prevention and control programs. Associated Activity B calls for improving regional coordination efforts. Michigan could support information sharing among states, especially on observed effects of climate change on ANS to improve prevention and management efforts.

B.14.4.2. Prevention

Objective 3 under the Legislative and Policy Goal calls for the development of a risk assessment process for potential and existing ANS. Activity C under this objective recommends developing a list of waters where additional assistance and/or effort could help reduce the spread of ANS. Because climate change will affect flows, water levels, dissolved oxygen content, and nutrient cycling, it will be important to consider how these changes may impact the susceptibility of high risk waters to ANS. Taking this information into consideration will help state staff more effectively target prevention activities.

B.14.4.3. Early Detection/Rapid Response, Control, and Management

Objective 4 under the Legislative and Policy Goal recommends the development of a regional ANS Rapid Response Team for areas with newly established ANS or areas threatened by potential introductions. Climate change may affect the ability of certain ANS to spread or become established; thus, it will be important to consider these effects when designing a rapid response plan.

B.14.4.4. Research

Activity F under Research and Monitoring Goal I calls for the development of a hot list of potential ANS that includes their locations, characteristics, and invasion probability. Any list of potential ANS should include climate change effects on the invasion pathway.

B.15.   MISSOURI AQUATIC NUISANCE SPECIES MANAGEMENT PLAN

B.15.1.   GENERAL DESCRIPTION OF MISSOURI'S PLAN

Missouri's Aquatic Nuisance Species (ANS) Management Plan was created by the Missouri Department of Conservation in 2005 (available at http://mdc.mo.gov/documents/nathis/exotic/ANSplan05.pdf). The Management Plan addresses three phases of invasion: introduction, spread, and abatement of impacts. The Plan outlines five goals with associated objectives and tasks to (1) inform the community about ANS and enlist their participation in halting introductions and spread; (2) collaborate on legislation development to prevent ANS; (3) monitor ANS distributions and conduct research to restrict spread; (4) develop and implement management to abate ANS impacts; and (5) abate ANS impacts. The Management Plan also includes an implementation table that identifies responsible agencies for each task and funding needs.

B.15.2.   CLIMATE CHANGE AND AQUATIC INVASIVE SPECIES IN MISSOURI

Temperatures in the Great Plains region are projected to rise as much as 3°F (1.5°C) in the summer and 4°F (2°C) in the winter by 2030. However, even these relatively small changes in temperature or precipitation (5 to 10% decline) could have significant effects on water quality, particularly salinity, and the availability of groundwater resources in the region (Covich et al., 1997). With higher temperatures and evaporation rates, stream and lake water levels may be lower in the summer. A large decrease in water levels could lead to shift in salinity and productivity in prairie pothole lakes and wetlands. In streams any increases in water temperatures, lowering of water levels, or increases in salinity will impact native fish species, such as the Plains killifish (*Fundulus zebrinus*) (Covich et al., 1997).

Warming water temperatures not only impact native fish, but also may enable certain ANS to survive and spread into areas where they currently cannot overwinter. For example, the Management Plan notes that water hyacinth cannot withstand the cold winters of Missouri. However, with warmer temperatures and the species' relatively wide temperature tolerance, water hyacinth could become a problem in Missouri. Dotted duckweed is already found in some parts of Missouri, but it could also expand its range if temperatures warm throughout the state.

B.15.3.   THE MISSOURI PLAN'S CURRENT INTEGRATION OF CLIMATE CHANGE

Table B-14 summarizes how the Missouri ANS Management Plan addresses and incorporates the predicted effects of climate change. Although Missouri's Plan does not explicitly account for climate change or changing conditions, it does recognize that species are influenced by climatic boundaries.

Table B-14. Assessment of the Missouri Aquatic Nuisance Species Management Plan

Aspects of plan that may incorporate climate change	Score
Understanding and incorporating potential impacts resulting from climate change 0 = no; 1 = briefly mentions; 2 = includes general discussion; 3 = includes quantitative info and/or specific examples	
Plan specifically mentions climate change	0
Plan acknowledges climatic boundaries of species	3
Plan demonstrates understanding of species and/or ecosystem sensitivity to changing conditions	0
Plan identifies research on the potential effects of species responding to changing conditions	0
Plan acknowledges regional differences in expected climate changes	0
Capacity to adapt to changing conditions 0 = no; 1 = implicitly (i.e., includes goals and strategies that can be used to account for changing conditions, but does not specify changing conditions as part of their purpose); 2 = yes, explicitly, in passing; 3 = yes, explicitly, and specifies associated goals and/or action items	
Plan accounts for changing conditions in its leadership and coordination goals and strategies	0
Plan accounts for changing conditions in its prevention goals and strategies	0
Plan accounts for changing conditions in its early detection/rapid response goals and strategies	0
Plan accounts for changing conditions in its control and management goals and strategies	0
Plan accounts for changing conditions in its restoration goals and strategies	0
Plan accounts for changing conditions in its research goals and strategies	0
Plan accounts for changing conditions in its information management goals and strategies	0
Plan accounts for changing conditions in its education and public awareness goals and strategies	0
Monitoring strategies 0 = no; 1 = yes, briefly mentions; 2 = yes, but unclear how information will be used; 3 = yes, and specifies associated goals and/or action items	
Plan includes strategy to monitor for changing conditions	0
Plan includes strategy to utilize monitoring data	3
Plan includes strategy for managing/updating monitoring data	3
Revision 0 = no; 1 = yes, in passing; 2 = yes, and includes qualitative description; 3 = yes, and includes timeline and/or benchmarks for doing so	
Plan includes strategy for updating and incorporating new information	0
Funding 0 = no; 1 = a source is specified for a portion of the required funding; 2 = a source is specified for a portion of the required funding along with strategies for obtaining remaining funding; 3 = a source is specified for 100% of required funding	
Plan identifies dedicated funding source for implementation	3
Total score:	**12**

B.15.4. INCORPORATING CLIMATE CHANGE INFORMATION

Given the potential effects of climate change in Missouri, various goals and objectives, and associated actions, may be less successful if climate change is not considered. Specifically, prevention and control efforts should consider climate change effects in order to better determine ANS threats and appropriate control methods.

B.15.4.1. Leadership and Coordination

Goals I and II address leadership, coordination, and communication about ANS issues. Climate change should be a consideration in communication on encroaching species and potential vectors and on adapting management practices to accommodate changing conditions. Objective IB calls for targeting education efforts at specific stakeholders and providing information on how ANS could harm resource of interest. Incorporating information on how ANS and climate change may interact and affect resources of interest, such as fisheries, could lead to a stronger response on the part of the public in regards to prevention efforts.

B.15.4.2. Prevention

Task IIA1 recommends continual review and revision of the state's Approved Aquatic Species List to have a baseline for evaluating which species can be safely brought into the state. The review should include an evaluation of how climate change affects approved and proposed species. Including this step may enable Missouri to maintain an effective prevention program.

B.15.4.3. Early Detection/Rapid Response, Control, and Management

Careful consideration should be given to how changing conditions could affect control methods when developing and implementing these strategies (Objective VB). Biocontrol methods could become less effective under warmer temperatures or biocontrol agents could become invasive as conditions change. Climate change may also affect ANS pathways. Warmer temperatures may increase recreational use of waterways. This could increase the possibility that species such as hydrilla and the New Zealand mud snail would be transported into or within the state by boats, or that the rusty crayfish would be introduced through bait releases.

B.15.4.4. Research

Objective IIIB calls for conducting and supporting research on ANS life histories, habitat use, potential effects on native species, and how they are transported and introduced into new areas. This research should also include an analysis of how climate change may affect each of these factors. For example, ANS effects on native species could be exacerbated by warmer waters that also damage a native species' habitat.

B.16. MONTANA AQUATIC NUISANCE SPECIES MANAGEMENT PLAN

B.16.1. GENERAL DESCRIPTION OF MONTANA'S PLAN

Montana's Aquatic Nuisance Species (ANS) Management Plan was written by the ANS Technical Committee (available at http://www.anstaskforce.gov/Montana-FINAL_PLAN.pdf). The goal of the Management Plan is to minimize ANS impacts through prevention and management of introduction, population growth, and dispersal. The goal is supported by six objectives to (1) coordinate and implement the Management Plan; (2) prevent ANS introductions; (3) detect, monitor, and eradicate new ANS; (4) control and eradicate established ANS; (5) communicate about ANS risks and impacts; and (6) increase and disseminate knowledge of ANS. The Management Plan includes implementation tables, expected funding, and monitoring and evaluation of the Plan's implementation.

B.16.2. CLIMATE CHANGE AND AQUATIC INVASIVE SPECIES IN MONTANA

Temperatures in the Great Plains region are projected to rise as much as 3°F (1.5°C) in the summer and 4°F (2°C) in the winter by 2030. However, even these relatively small changes in temperature or precipitation (5 to 10% decline) could have significant effects on water quality, particularly salinity, and the availability of groundwater resources in the region (Covich et al., 1997). With higher temperatures and evaporation rates, stream and lake water levels may be lower in the summer. The warmer climate also could cause earlier snowmelt in the spring, resulting in higher stream flows in the winter and spring and lower ones in the summer and fall. A large decrease in water levels could lead to shift in salinity and productivity in prairie pothole lakes and wetlands.

Increased water temperatures can be an added stressor for fish, potentially resulting in increased fish mortality from non-native bacterial fish pathogens, such as *Aeromonas salmonicida*, which already are present in some Montana watersheds. The decline of native salmonids may facilitate the spread of non-native fish such as bass and walleye (already present in the state and used for sport fishing, but still under careful management to prevent spread).

B.16.3. THE MONTANA PLAN'S CURRENT INTEGRATION OF CLIMATE CHANGE

Table B-15 summarizes how the Montana ANS Management Plan addresses and incorporates the projected effects of climate change. The Management Plan does not specifically address the effects of climate change on its management objectives, but some elements in the Plan allow for changing conditions to be considered in implementation of the Plan.

Table B-15. Assessment of the Montana Aquatic Nuisance Species Management Plan

Aspects of plan that may incorporate climate change	Score
Understanding and incorporating potential impacts resulting from climate change 0 = no; 1 = briefly mentions; 2 = includes general discussion; 3 = includes quantitative info and/or specific examples	
Plan specifically mentions climate change	0
Plan acknowledges climatic boundaries of species	0
Plan demonstrates understanding of species and/or ecosystem sensitivity to changing conditions	1
Plan identifies research on the potential effects of species responding to changing conditions	0
Plan acknowledges regional differences in expected climate changes	0
Capacity to adapt to changing conditions 0 = no; 1 = implicitly (i.e., includes goals and strategies that can be used to account for changing conditions, but does not specify changing conditions as part of their purpose); 2 = yes, explicitly, in passing; 3 = yes, explicitly, and specifies associated goals and/or action items	
Plan accounts for changing conditions in its leadership and coordination goals and strategies	0
Plan accounts for changing conditions in its prevention goals and strategies	0
Plan accounts for changing conditions in its early detection/rapid response goals and strategies	0
Plan accounts for changing conditions in its control and management goals and strategies	0
Plan accounts for changing conditions in its restoration goals and strategies	0
Plan accounts for changing conditions in its research goals and strategies	3
Plan accounts for changing conditions in its information management goals and strategies	0
Plan accounts for changing conditions in its education and public awareness goals and strategies	0
Monitoring strategies 0 = no; 1 = yes, briefly mentions; 2 = yes, but unclear how information will be used; 3 = yes, and specifies associated goals and/or action items	
Plan includes strategy to monitor for changing conditions	0
Plan includes strategy to utilize monitoring data	0
Plan includes strategy for managing/updating monitoring data	0
Revision 0 = no; 1 = yes, in passing; 2 = yes, and includes qualitative description; 3 = yes, and includes timeline and/or benchmarks for doing so	
Plan includes strategy for updating and incorporating new information	3
Funding 0 = no; 1 = a source is specified for a portion of the required funding; 2 = a source is specified for a portion of the required funding along with strategies for obtaining remaining funding; 3 = a source is specified for 100% of required funding	
Plan identifies dedicated funding source for implementation	2
Total score:	**9**

B.16.4. INCORPORATING CLIMATE CHANGE INFORMATION

Given the effects due to climate change projected for Montana, state staff might consider incorporating information on these impacts into both implementation of the current Management Plan and subsequent revisions to the Plan.

B.16.4.1. Leadership and Coordination

Strategy 1B calls for participation in and support of regional, federal, and international efforts to control ANS. In order to better anticipate and manage the expanding ranges of ANS, regional efforts should be cognizant of and in communication about species' expanding ranges.

B.16.4.2. Prevention

Strategy 2A establishes the task of describing invasion pathways and identifying high-risk water bodies. Analysis of invasion pathways should account for climate-sensitive vectors, such as increased recreational boating and the impact of increased water temperature on the health of native fish and the ability of non-native fish to establish new populations.

B.16.4.3. Early Detection/Rapid Response, Control, and Management

Identification of high-risk water bodies might reflect the impacts on native and invasive species of changes such as reduced lake water levels and decreased stream flows caused by increased evaporation. Strategy 2C, designed to prohibit, control, or permit the importation of non-indigenous aquatic species based upon their invasive potential, requires research on the invasive potential of aquatic plant species currently imported. This research should examine currently permissible species' ability to persist in the expected conditions resulting from climate change in order to obtain a complete assessment of invasive potential.

B.16.4.4. Research

Strategy 6B, which relates to research on management alternatives and their effects on ANS and native species, calls for the investigation of the relationship between human-induced disturbance of aquatic and riparian systems and ANS invasion, establishment, and impacts. This is an ideal opportunity for state staff to research the effects of climate change on the state's water bodies and waterways and the influence of these changes on ANS invasions.

B.17. NEW YORK NONINDIGENOUS AQUATIC SPECIES COMPREHENSIVE MANAGEMENT PLAN

B.17.1. GENERAL DESCRIPTION OF NEW YORK'S PLAN

New York's Nonindigenous Aquatic Species (NAS) Comprehensive Management Plan was written by the New York State Department of Environmental Conservation in 1993 (available at http://www.anstaskforce.gov/State%20Plans/NY%20Mgt%20Plan%201993.pdf). The goals of the Management Plan are to prevent NAS from being introduced, to limit the spread and impacts of NAS already present, and to educate the public on preventing introductions and reducing impacts. New York's Management Plan outlines the problems related to accomplishing the goals and related actions to overcome these problems. The Plan also includes recommendations for implementation and an implementation schedule.

B.17.2. CLIMATE CHANGE AND AQUATIC INVASIVE SPECIES IN NEW YORK

Projected increases in annual surface temperatures in the Northeastern region of the United States are projected to average 10°F (5.3°C) by 2070. Nearly all model simulations of future precipitation show consistent increases in winter precipitation and no change to a decrease in summer rainfall. By 2100, precipitation is projected to increase an average of 11 to 14% in the winter. Regional sea surface temperatures are projected to increase in accordance with regional air temperatures; these increasing temperatures have the potential to expand the range of warm-water species northward and permit invasive species to spread into these waters, which had previously been previously too cold to allow for invasive species' survival (Hayhoe et al., 2007).

Evaporation is likely to increase with warmer temperatures, resulting in lower river flow and lower lake water levels in the summer and fall. In general, New York is expected to experience higher stream flow in the winter and spring and lower stream flow in the summer and fall (Hayhoe et al., 2007). This may give a competitive advantage to the round goby, a problematic aquatic invasive species in New York that can survive in degraded water conditions (USACE, 2007).

B.17.3. THE NEW YORK PLAN'S CURRENT INTEGRATION OF CLIMATE CHANGE

Table B-16 summarizes how the NAS Comprehensive Management Plan addresses and incorporates the projected effects of climate change. Although the Management Plan does not incorporate climate change, it has strong provisions for monitoring new introductions of NAS.

Table B-16. Assessment of the New York Aquatic Nuisance Species Management Plan

Aspects of plan that may incorporate climate change	Score
Understanding and incorporating potential impacts resulting from climate change 0 = no; 1 = briefly mentions; 2 = includes general discussion; 3 = includes quantitative info and/or specific examples	
Plan specifically mentions climate change	0
Plan acknowledges climatic boundaries of species	0
Plan demonstrates understanding of species and/or ecosystem sensitivity to changing conditions	0
Plan identifies research on the potential effects of species responding to changing conditions	0
Plan acknowledges regional differences in expected climate changes	0
Capacity to adapt to changing conditions 0 = no; 1 = implicitly (i.e., includes goals and strategies that can be used to account for changing conditions, but does not specify changing conditions as part of their purpose); 2 = yes, explicitly, in passing; 3 = yes, explicitly, and specifies associated goals and/or action items	
Plan accounts for changing conditions in its leadership and coordination goals and strategies	0
Plan accounts for changing conditions in its prevention goals and strategies	0
Plan accounts for changing conditions in its early detection/rapid response goals and strategies	0
Plan accounts for changing conditions in its control and management goals and strategies	0
Plan accounts for changing conditions in its restoration goals and strategies	0
Plan accounts for changing conditions in its research goals and strategies	0
Plan accounts for changing conditions in its information management goals and strategies	0
Plan accounts for changing conditions in its education and public awareness goals and strategies	0
Monitoring strategies 0 = no; 1 = yes, briefly mentions; 2 = yes, but unclear how information will be used; 3 = yes, and specifies associated goals and/or action items	
Plan includes strategy to monitor for changing conditions	0
Plan includes strategy to utilize monitoring data	3
Plan includes strategy for managing/updating monitoring data	0
Revision 0 = no; 1 = yes, in passing; 2 = yes, and includes qualitative description; 3 = yes, and includes timeline and/or benchmarks for doing so	
Plan includes strategy for updating and incorporating new information	0
Funding 0 = no; 1 = a source is specified for a portion of the required funding; 2 = a source is specified for a portion of the required funding along with strategies for obtaining remaining funding; 3 = a source is specified for 100% of required funding	
Plan identifies dedicated funding source for implementation	1
Total score:	**4**

B.17.4. INCORPORATING CLIMATE CHANGE INFORMATION

Given the projected effects of climate change for New York, consideration of these effects in the Management Plan could increase the efficacy of NAS prevention and control. Several goals and strategies outlined in the Plan could incorporate climate change considerations.

B.17.4.1. Leadership and Coordination

Goal 4 addresses communication with the public and coordinating with relevant groups. Climate change should be a consideration in communication on encroaching species and potential vectors and on adapting management practices to accommodate changing conditions.

B.17.4.2. Prevention

Goal 1 calls for the identification of aquatic organisms that could potentially have adverse impacts in state waters, as well as the characteristics, habitat requirements, and potential adverse impacts of these organisms. Assessment of adverse impacts should take into account the changing conditions projected by climate change models, which may make systems more vulnerable and/or exacerbate the effects of NAS. For example, warmer water temperatures could negatively impact native trout fisheries, allowing both NAS to become more easily established and greater impacts on the ecosystem.

B.17.4.3. Early Detection/Rapid Response, Control, and Management

Climate change considerations could also be incorporated into the Plan to strengthen progress toward Goal 2, reducing the potential for NAS that have been introduced into state waters to spread to uncolonized waters. This goal calls for monitoring colonized waters and collecting data on NAS such as rate of growth, distribution, and impacts on native species. These data are to be correlated with habitat data to develop predictive models of where and how NAS introductions might occur and to develop strategies for preventing and controlling them. These models could incorporate predicted habitat changes that may result from climate change to better predict NAS introductions.

B.17.4.4. Research

The aim of Goal 3 is to provide information and management strategies related to minimizing impacts from NAS. Development of alternative management strategies and risk assessments of proposed introductions should consider climate change effects to ensure long-term efficacy.

B.18.   NORTH DAKOTA AQUATIC NUISANCE SPECIES MANAGEMENT PLAN

B.18.1.   GENERAL DESCRIPTION OF NORTH DAKOTA'S PLAN

North Dakota's Aquatic Nuisance Species (ANS) Management Plan was written by the North Dakota Game and Fish Department (available at http://gf.nd.gov/fishing/docs/nd-ans-plan-fnl-drft.pdf). The goal of the Management Plan is to prevent ANS impacts through seven supporting objectives. The objectives are to (1) coordinate ANS activities, (2) prevent introductions, (3) detect new ANS and monitor existing populations, (4) educate the public to prevent ANS spread, (5) control and eradicate established ANS, (6) inform policy makers about ANS risks and impacts, and (7) increase ANS knowledge base. North Dakota's Management Plan includes a budget and time frame, regulatory needs, and prioritized Strategic Actions.

B.18.2.   CLIMATE CHANGE AND AQUATIC INVASIVE SPECIES IN NORTH DAKOTA

Temperatures in the Great Plains region are projected to rise as much as 3°F (1.5°C) in the summer and 4°F (2°C) in the winter by 2030. However, even these relatively small changes in temperature or precipitation (5 to 10% decline) could have significant effects on water quality, particularly salinity, and groundwater availability (Covich et al., 1997). With higher temperatures and evaporation rates, water levels may be lower in the summer, particularly in prairie pothole lakes and wetlands. The warmer climate also could cause earlier snowmelt, resulting in higher stream flows in the winter and spring and lower ones in the summer and fall. Increases in water temperatures, lower water levels, or increases in salinity may impact native fish (Covich et al., 1997).

The potentially significant changes in hydrology associated with rising temperatures could make North Dakota's native species and habitats more vulnerable to invasive species. For example, salt cedar, an aquatic invasive species of concern already established in North Dakota, is tolerant of dry conditions and could become a larger problem under climate change conditions as it out-competes native plants for water. Additionally, North Dakota's Management Plan includes Whirling disease (*Myxobolus cerebralis*) as a significant invasive threat, which may become more prevalent as a result of climate change, impacting salmon populations.

B.18.3.   THE NORTH DAKOTA PLAN'S CURRENT INTEGRATION OF CLIMATE CHANGE

Table B-17 summarizes how the North Dakota ANS Management Plan addresses and incorporates the projected effects of climate change. The Management Plan does not specifically address climate change in its objectives, but does describe management actions that are intended to be adaptable to changing circumstances. The Plan also recognizes that certain ANS currently cannot survive in North Dakota due to cold temperatures.

Table B-17. Assessment of the North Dakota Aquatic Nuisance Species Management Plan

Aspects of plan that may incorporate climate change	Score
Understanding and incorporating potential impacts resulting from climate change 0 = no; 1 = briefly mentions; 2 = includes general discussion; 3 = includes quantitative info and/or specific examples	
Plan specifically mentions climate change	0
Plan acknowledges climatic boundaries of species	2
Plan demonstrates understanding of species and/or ecosystem sensitivity to changing conditions	1
Plan identifies research on the potential effects of species responding to changing conditions	0
Plan acknowledges regional differences in expected climate changes	0
Capacity to adapt to changing conditions 0 = no; 1 = implicitly (i.e., includes goals and strategies that can be used to account for changing conditions, but does not specify changing conditions as part of their purpose); 2 = yes, explicitly, in passing; 3 = yes, explicitly, and specifies associated goals and/or action items	
Plan accounts for changing conditions in its leadership and coordination goals and strategies	0
Plan accounts for changing conditions in its prevention goals and strategies	0
Plan accounts for changing conditions in its early detection/rapid response goals and strategies	0
Plan accounts for changing conditions in its control and management goals and strategies	0
Plan accounts for changing conditions in its restoration goals and strategies	0
Plan accounts for changing conditions in its research goals and strategies	1
Plan accounts for changing conditions in its information management goals and strategies	0
Plan accounts for changing conditions in its education and public awareness goals and strategies	0
Monitoring strategies 0 = no; 1 = yes, briefly mentions; 2 = yes, but unclear how information will be used; 3 = yes, and specifies associated goals and/or action items	
Plan includes strategy to monitor for changing conditions	0
Plan includes strategy to utilize monitoring data	1
Plan includes strategy for managing/updating monitoring data	1
Revision 0 = no; 1 = yes, in passing; 2 = yes, and includes qualitative description; 3 = yes, and includes timeline and/or benchmarks for doing so	
Plan includes strategy for updating and incorporating new information	1
Funding 0 = no; 1 = a source is specified for a portion of the required funding; 2 = a source is specified for a portion of the required funding along with strategies for obtaining remaining funding; 3 = a source is specified for 100% of required funding	
Plan identifies dedicated funding source for implementation	2
Total score:	**9**

B.18.4. INCORPORATING CLIMATE CHANGE INFORMATION

Given the expected impacts of climate change in North Dakota, state staff should consider climate change predictions in its Plan objectives, actions, and strategies to increase the effectiveness of prevention and management efforts. For example, state staff could specifically include predicted conditions resulting from climate change when determining ANS invasive potential.

B.18.4.1. Leadership and Coordination

Objective 1 addresses coordination about ANS issues. Climate change should be a consideration in communication on encroaching species and potential vectors and on adapting management practices to accommodate changing conditions.

B.18.4.2. Prevention

A primary objective of the Management Plan is to prevent ANS. Associated actions call for creating a list of potential problem ANS. Additionally, Strategy 2A2 calls for conducting risk analyses for each potential ANS introduction pathway. In determining potential ANS, research should include species' ability to survive in the expected conditions resulting from climate change in order to obtain a more complete assessment of invasive potential. Assessments should include responses to projected effects such as increased flooding, decreased stream flows, and increased water temperatures. For example, ANS introductions from the aquarium trade are a potential problem North Dakota; thus, assessments could address the aquarium fish species that are expected to thrive under the predicted temperature increases for North Dakota's waters.

B.18.4.3. Early Detection/Rapid Response, Control, and Management

Strategy 3A2 recommends conducting annual monitoring of high risk waters as a part of implementing a monitoring and early detection program. Considering climate change in these assessments may help state staff determine which waters are most vulnerable to species invasions.

B.18.4.4. Research

Objective 5 calls for controlling and eradicating pioneering and established ANS. State staff should conduct research to determine how different management strategies could be impacted by temperature, precipitation, and water level changes, and state staff should consider climate change impacts on control and management methods in developing management plans.

B.19. OHIO COMPREHENSIVE MANAGEMENT PLAN FOR AQUATIC NUISANCE SPECIES

B.19.1. GENERAL DESCRIPTION OF OHIO'S PLAN

Ohio's Department of Natural Resources led the writing team that developed the Comprehensive Management Plan for Aquatic Nuisance Species (ANS) (available at http://www.anstaskforce.gov/State%20Plans/Ohio%20Comprehensive%20Management%20Plan .pdf). The Management Plan provides a framework for the state's future approach to ANS management and builds support for work plans in development and funding requests. The goal of the Management Plan is to decrease the rate of introductions that have increased since the St. Lawrence Seaway was established in 1960 and transportation levels in the Great Lakes sky-rocketed. The Management Plan provides guidance on developing management actions to prevent, control, and limit the impacts of established and potential ANS in Ohio's inland waters.

B.19.2. CLIMATE CHANGE AND AQUATIC INVASIVE SPECIES IN OHIO

Temperatures are projected to increase by 5 to10°F (2.8 to 5.5°C) in the northern portion of Midwest region throughout the 21st century. Precipitation is expected to increase by 10 to 30% across the region. Increasing temperatures can increase evaporation, causing soil moisture deficits, reduction in lake and river water levels, and more drought-like conditions. For smaller lakes and rivers, reduced flows are likely to intensify water quality issues. Eutrophication of lakes may increase due to increases in excess nutrient runoff from heavy precipitation events and warmer lake temperatures that stimulate algae growth (Easterling and Karl, 2001).

As water temperatures in lakes increase, significant changes in freshwater ecosystems will occur, such as a shift from cold water fish species, such as trout, to warmer water species, like bass and catfish. Warmer water temperatures would create an environment that is more susceptible to invasions by non-native species (Easterling and Karl, 2001). If lake water levels decrease, more shoreline could become exposed, making this area vulnerable to purple loosestrife infestation, which destroys native habitats and can impact bird species.

B.19.3. THE OHIO PLAN'S CURRENT INTEGRATION OF CLIMATE CHANGE

Table B-18 summarizes how the Ohio Comprehensive Management Plan addresses and incorporates the projected effects of climate change. The Management Plan details Lake Erie's vulnerability to both habitat and trophic changes, as well as how this can lead to more invasive species. Although the Management Plan does not contain an implementation plan, it does have a framework that could be used to manage waters under a changing climate.

Table B-18. Assessment of the Ohio Aquatic Nuisance Species Management Plan

Aspects of plan that may incorporate climate change	Score
Understanding and incorporating potential impacts resulting from climate change 0 = no; 1 = briefly mentions; 2 = includes general discussion; 3 = includes quantitative info and/or specific examples	
Plan specifically mentions climate change	0
Plan acknowledges climatic boundaries of species	1
Plan demonstrates understanding of species and/or ecosystem sensitivity to changing conditions	0
Plan identifies research on the potential effects of species responding to changing conditions	0
Plan acknowledges regional differences in expected climate changes	0
Capacity to adapt to changing conditions 0 = no; 1 = implicitly (i.e., includes goals and strategies that can be used to account for changing conditions, but does not specify changing conditions as part of their purpose); 2 = yes, explicitly, in passing; 3 = yes, explicitly, and specifies associated goals and/or action items	
Plan accounts for changing conditions in its leadership and coordination goals and strategies	0
Plan accounts for changing conditions in its prevention goals and strategies	0
Plan accounts for changing conditions in its early detection/rapid response goals and strategies	0
Plan accounts for changing conditions in its control and management goals and strategies	0
Plan accounts for changing conditions in its restoration goals and strategies	0
Plan accounts for changing conditions in its research goals and strategies	2
Plan accounts for changing conditions in its information management goals and strategies	0
Plan accounts for changing conditions in its education and public awareness goals and strategies	0
Monitoring strategies 0 = no; 1 = yes, briefly mentions; 2 = yes, but unclear how information will be used; 3 = yes, and specifies associated goals and/or action items	
Plan includes strategy to monitor for changing conditions	0
Plan includes strategy to utilize monitoring data	1
Plan includes strategy for managing/updating monitoring data	0
Revision 0 = no; 1 = yes, in passing; 2 = yes, and includes qualitative description; 3 = yes, and includes timeline and/or benchmarks for doing so	
Plan includes strategy for updating and incorporating new information	1
Funding 0 = no; 1 = a source is specified for a portion of the required funding; 2 = a source is specified for a portion of the required funding along with strategies for obtaining remaining funding; 3 = a source is specified for 100% of required funding	
Plan identifies dedicated funding source for implementation	0
Total score:	**5**

B.19.4. INCORPORATING CLIMATE CHANGE INFORMATION

Ohio is vulnerable to ANS, as well as to the effects of climate change. Prevention and monitoring efforts should consider climate change effects in order to better determine ANS threats and ecosystem vulnerabilities.

B.19.4.1. Leadership and Coordination

Ohio has established a framework for successful ANS management that considers many of the necessary contributing factors to ANS spread. It also calls for coordinating prevention and control strategies. However, the data needed to help guide management actions within the context of climate change are lacking. A monitoring program that incorporates climate change considerations and uses this information to help identify areas of Lake Erie and inland lakes that are especially vulnerable to climate change will help the state properly allocate resources and implement management strategies effectively.

B.19.4.2. Prevention

State staff is aware of Ohio's sensitive position within the Great Lakes region, and the state is proactive in listing transport mechanisms that require regional attention in the Plan's management actions section. For example, Ohio highlights transport mechanisms such as the flushing of natural gas pipelines from northerly regions with water as a potential ANS pathway that has not been fully documented. As climate change alters nutrient cycles and ecosystems, aquatic habitats may become more vulnerable to ANS arriving through these pathways.

B.19.4.3. Early Detection/Rapid Response, Control, and Management

As the Management Plan notes, Lake Erie is the most vulnerable water body to ANS and a potential vector for ANS. Any established species in this area must be kept from spreading to the rest of the state and to the Ohio River to control its widespread dispersal. Monitoring how habitats are affected by climate change can help target control strategies and ensure they are successful.

B.19.4.4. Research

Strategic Action IA calls for researching the movement and transport of ANS on a global scale. Incorporating climate change considerations into these analyses can help state staff better predict potential introductions and effectively target prevention strategies.

B.20. OREGON AQUATIC NUISANCE SPECIES MANAGEMENT PLAN

B.20.1. GENERAL DESCRIPTION OF OREGON'S PLAN

Oregon's Aquatic Nuisance Species (ANS) Management Plan was prepared in 2001 by the Center for Lakes and Reservoirs at Portland State University with direction and participation from state staff (available at http://www.clr.pdx.edu/publications/files/OR_ANS_Plan.pdf). The goals of the Management Plan are to prevent and manage ANS introductions, population growth, and spread to reduce impacts throughout the state. In support of these goals the Management Plan outlines six objectives to (1) coordinate and implement the Plan; (2) prevent introductions; (3) detect, monitor, and eradicate new ANS; (4) control established ANS; (5) inform about ANS risks and impacts; and (6) increase and disseminate knowledge about ANS. The Management Plan includes an implementation table with a corresponding budget.

B.20.2. CLIMATE CHANGE AND AQUATIC INVASIVE SPECIES IN OREGON

Average warming in the Pacific Northwest is projected to increase by 3°F (1.7°C) by the 2020s and 5°F (2.8°C) by the 2050s. Annual precipitation projections are less certain, ranging from 7% or 2 inches (5cm) to a 13% or 4 inches (10cm). Heavier winter rainfall could increase soil saturation, landslides, and winter flooding. In addition, projected increases in mean sea level may increase sediment erosion and redistribution on the open coast (Parson, 2001b). Projected precipitation increases will be concentrated in winter, with decreases (or smaller increases) in summer; for this reason, even the projections that show increases in annual precipitation show decreases in water availability (Parson, 2001b).

Various invasive aquatic plants identified in the Plan, such as giant salvinia and hydrilla, destroy native fish habitat and alter water chemistry. These impacts will interact with warmer water temperatures that also damage and reduce fish habitat. Additionally, as water levels decrease in the summer, less water is available for domestic and agricultural purposes, as well as for hydropower. Species such as zebra mussels may this problem by blocking intake pipes.

B.20.3. THE OREGON PLAN'S CURRENT INTEGRATION OF CLIMATE CHANGE

Table B-19 summarizes how the Oregon ANS Management Plan addresses and incorporates the projected effects of climate change. The ANS Management Plan does not explicitly mention climate change; however, the Plan does convey that some ANS of concern may not currently be a major threat to Oregon because of their climate tolerance. Additionally, updates and revisions to the Plan should incorporate changing circumstances.

Table B-19. Assessment of the Oregon Aquatic Nuisance Species Management Plan

Aspects of plan that may incorporate climate change	Score
Understanding and incorporating potential impacts resulting from climate change 0 = no; 1 = briefly mentions; 2 = includes general discussion; 3 = includes quantitative info and/or specific examples	
Plan specifically mentions climate change	0
Plan acknowledges climatic boundaries of species	3
Plan demonstrates understanding of species and/or ecosystem sensitivity to changing conditions	0
Plan identifies research on the potential effects of species responding to changing conditions	0
Plan acknowledges regional differences in expected climate changes	0
Capacity to adapt to changing conditions 0 = no; 1 = implicitly (i.e., includes goals and strategies that can be used to account for changing conditions, but does not specify changing conditions as part of their purpose); 2 = yes, explicitly, in passing; 3 = yes, explicitly, and specifies associated goals and/or action items	
Plan accounts for changing conditions in its leadership and coordination goals and strategies	0
Plan accounts for changing conditions in its prevention goals and strategies	0
Plan accounts for changing conditions in its early detection/rapid response goals and strategies	0
Plan accounts for changing conditions in its control and management goals and strategies	0
Plan accounts for changing conditions in its restoration goals and strategies	0
Plan accounts for changing conditions in its research goals and strategies	0
Plan accounts for changing conditions in its information management goals and strategies	0
Plan accounts for changing conditions in its education and public awareness goals and strategies	0
Monitoring strategies 0 = no; 1 = yes, briefly mentions; 2 = yes, but unclear how information will be used; 3 = yes, and specifies associated goals and/or action items	
Plan includes strategy to monitor for changing conditions	0
Plan includes strategy to utilize monitoring data	0
Plan includes strategy for managing/updating monitoring data	0
Revision 0 = no; 1 = yes, in passing; 2 = yes, and includes qualitative description; 3 = yes, and includes timeline and/or benchmarks for doing so	
Plan includes strategy for updating and incorporating new information	3
Funding 0 = no; 1 = a source is specified for a portion of the required funding; 2 = a source is specified for a portion of the required funding along with strategies for obtaining remaining funding; 3 = a source is specified for 100% of required funding	
Plan identifies dedicated funding source for implementation	3
Total score:	**9**

B.20.4. INCORPORATING CLIMATE CHANGE INFORMATION

Given the projected effects of climate change for Oregon, consideration of these effects in the Management Plan could increase the efficacy of ANS prevention and control. Several goals and strategies outlined in the Plan could incorporate climate change considerations.

B.20.4.1. Leadership and Coordination

Strategy 1A calls for coordination between all ANS management programs and activities within Oregon, which could incorporate climate change considerations. For example, the annual symposium should discuss how management alternatives could incorporate climate change effects. Climate change information may also be incorporated into ANS assessment guidelines and a prioritization scheme. For example, a species' ability to become established or spread throughout the state may be affected by water temperatures, water chemistry, and water levels; thus, how climate change will affect these factors should be considered in these processes.

B.20.4.2. Prevention

Strategic Action 2A2 recommends describing introduction pathways and identifying high-risk water bodies. In evaluating high-risk water bodies, climate change considerations should be incorporated into the analysis. Water bodies that are currently unaffected by ANS due to cooler temperatures may become vulnerable in the future as water temperature increases. Integrating these considerations into the identification process may improve prevention efforts.

B.20.4.3. Early Detection/Rapid Response, Control, and Management

Strategy 3A calls for the implementation of a surveillance and early detection program. Monitoring programs should incorporate climate change considerations in order to target areas and water bodies that may become vulnerable to invasion. For example, the Plan identifies giant salvinia as a potential threat to Oregon's waters. This plant is primarily transported by humans, especially recreational boat users. Based on the plant's biology, it should be able to survive climatic conditions in western Oregon. However, in the face of climate change, this species could perhaps become invasive in other regions as well. A monitoring program designed to detect changes in ecosystem vulnerabilities will remain more effective at detecting new ANS.

B.20.4.4. Research

Strategy 6A calls for research in ANS impacts on native species. Incorporating climate change considerations into research will improve impact assessments and the development of management strategies that remain effective under changing conditions.

B.21. PENNSYLVANIA AQUATIC INVASIVE SPECIES MANAGEMENT PLAN

B.21.1. GENERAL DESCRIPTION OF PENNSYLVANIA'S PLAN

The Pennsylvania Aquatic Invasive Species (AIS) Management Plan was written by the AIS Management Plan Committee (AISMPC) for the Pennsylvania Invasive Species Council (available at http://www.holstoncrisci.com/Newsletter/docs/3/PAAISMP.pdf). The goals of the Management Plan are to prevent and manage the introduction and spread of AIS to minimize negative impacts. Eight objectives to meet these goals include (1) providing leadership and coordination; (2) identifying and minimizing vectors of introductions; (3) detecting introductions; (4) developing early detection/rapid response; (5) monitoring existing AIS; (6) controlling and eradicating established AIS; (7) increasing AIS research; and (8) educating people about AIS issues. Objectives have related strategies and actions, which are prioritized. The Plan also includes evaluation and revision strategies and an implementation table.

B.21.2. CLIMATE CHANGE AND AQUATIC INVASIVE SPECIES IN PENNSYLVANIA

Projected increases in annual surface temperatures in the Northeastern region of the United States are projected to average 10°F (5.3°C) by 2070. Nearly all model simulations of future precipitation show consistent increases in winter precipitation and no change to a decrease in summer rainfall. By 2100, precipitation is projected to increase an average of 11 to 14% in the winter (Hayhoe et al., 2007).

Warming air and water temperatures, water level fluctuations, and water chemistry changes could affect AIS establishment and spread. According to the Plan, hydrilla currently is found in three water bodies in Pennsylvania. As temperatures rise, recreation activities may become more widespread and prolonged, which may result in more introductions of hydrilla into other parts of the state. Hydrilla forms thick mats that can block intake pipes, which could hinder the provision of water. If climate change results in larger stream flows and altered water chemistry, the European rudd fish may out-compete native species as it is a hardy species that can survive in polluted waters.

B.21.3. THE PENNSYLVANIA PLAN'S CURRENT INTEGRATION OF CLIMATE CHANGE

Table B-20 summarizes how the Pennsylvania AIS Management Plan addresses and incorporates the projected effects of climate change. Although the Management Plan does not specifically address climate change, it has strong provisions for developing and conducting risk analyses to prevent introductions of AIS.

Table B-20. Assessment of the Pennsylvania Aquatic Nuisance Species Management Plan

Aspects of plan that may incorporate climate change	Score
Understanding and incorporating potential impacts resulting from climate change 0 = no; 1 = briefly mentions; 2 = includes general discussion; 3 = includes quantitative info and/or specific examples	
Plan specifically mentions climate change	0
Plan acknowledges climatic boundaries of species	0
Plan demonstrates understanding of species and/or ecosystem sensitivity to changing conditions	0
Plan identifies research on the potential effects of species responding to changing conditions	0
Plan acknowledges regional differences in expected climate changes	0
Capacity to adapt to changing conditions 0 = no; 1 = implicitly (i.e., includes goals and strategies that can be used to account for changing conditions, but does not specify changing conditions as part of their purpose); 2 = yes, explicitly, in passing; 3 = yes, explicitly, and specifies associated goals and/or action items	
Plan accounts for changing conditions in its leadership and coordination goals and strategies	0
Plan accounts for changing conditions in its prevention goals and strategies	0
Plan accounts for changing conditions in its early detection/rapid response goals and strategies	0
Plan accounts for changing conditions in its control and management goals and strategies	0
Plan accounts for changing conditions in its restoration goals and strategies	0
Plan accounts for changing conditions in its research goals and strategies	0
Plan accounts for changing conditions in its information management goals and strategies	0
Plan accounts for changing conditions in its education and public awareness goals and strategies	0
Monitoring strategies 0 = no; 1 = yes, briefly mentions; 2 = yes, but unclear how information will be used; 3 = yes, and specifies associated goals and/or action items	
Plan includes strategy to monitor for changing conditions	0
Plan includes strategy to utilize monitoring data	0
Plan includes strategy for managing/updating monitoring data	0
Revision 0 = no; 1 = yes, in passing; 2 = yes, and includes qualitative description; 3 = yes, and includes timeline and/or benchmarks for doing so	
Plan includes strategy for updating and incorporating new information	2
Funding 0 = no; 1 = a source is specified for a portion of the required funding; 2 = a source is specified for a portion of the required funding along with strategies for obtaining remaining funding; 3 = a source is specified for 100% of required funding	
Plan identifies dedicated funding source for implementation	1
Total score:	**3**

B.21.4. INCORPORATING CLIMATE CHANGE INFORMATION

Given Pennsylvania's sensitivity to climate change impacts and the range of potential and already established AIS, it is important for state staff to incorporate climate change considerations into the next version of their Management Plan. Climate change concerns could be integrated into several strategies and actions of the current Management Plan.

B.21.4.1. Leadership and Coordination

Strategic Action IC1 calls for partnering with states in the region to share data and coordinate management activities to prevent AIS introductions. By collecting information from adjacent and southern states, state staff can better determine what species may spreading in response to climate changes and be better prepared to implement a rapid response program.

B.21.4.2. Prevention

Strategy 2A and associated Strategic Action 2A1 call for establishing a comprehensive process to identify AIS of concern using scientific methods and research-based risk analysis. Climate change data should be incorporated into these analyses. For example, AIS that currently are not established in Pennsylvania because they cannot overwinter may establish if temperatures rise. Additionally, recreational boat use and fishing may increase with warmer temperatures, which could lead to increased AIS transport opportunities.

B.21.4.3. Early Detection/Rapid Response, Control, and Management

Strategic Action 6B1 calls for cost-benefit analyses to prioritize AIS control activities for species and sites. Because control strategies can be sensitive to environmental conditions, cost-benefit analyses should incorporate an evaluation of how warmer water temperatures or water availability changes could influence the effectiveness and cost of control strategies.

B.21.4.4. Research

Strategy 7A calls for establishing and coordinating an AIS research network that can help prioritize research needs and research and summarize effective control and management actions. Incorporating climate change considerations into these research efforts can inform prioritizations of AIS research needs and improve the development of management strategies that remain effective under changing conditions.

B.22. SOUTH CAROLINA AQUATIC PLANT MANAGEMENT PLAN PART I AND II

B.22.1. GENERAL DESCRIPTION OF SOUTH CAROLINA'S PLAN

South Carolina's Department of Natural Resources in cooperation with the state's Aquatic Plant Management Council develops an Aquatic Plant Management Plan each year (available at http://www.dnr.sc.gov/water/envaff/aquatic/plan.html). The purpose of the 2006 Management Plan is to protect the state's public waters from the adverse effects of aquatic plant populations. The 2006 Plan identifies five actions to (1) identify existing and potential aquatic plan problems; (2) determine appropriate control strategies; (3) develop operational strategies to implement control; (4) seek funding for implementation; and (5) monitor results and determine needs for modification. The Management Plan describes problems and specific control strategies actions for various aquatic nuisance plants to be implemented by water body.

B.22.2. CLIMATE CHANGE AND AQUATIC INVASIVE SPECIES IN SOUTH CAROLINA

Climate models project increases in annual summer air temperatures of 5.5°F (3°C) and winter air temperatures by 7°F (4°C) in the southeast. Climate model results are less certain for precipitation, but general indications are that there may be a 10% increase in summer precipitation and a 5% increase in winter precipitation. While sea levels are expected to rise, the rate of increase is highly uncertain. Sea level rise could result in significant coastal wetland loss, increasing open water areas and estuarine depths (Mulholland et al., 1997). The loss of coastal wetlands and marshes has the potential to reduce estuarine productivity because many estuarine species depend on wetlands as nursery areas and sources of organic matter (Mulholland et al., 1997).

These climate change effects may also influence aquatic invasive species (AIS). For example, the Bear Island Wilderness Management Area is a low-lying coastal wetland currently threatened by several aquatic plant species, including cutgrass and phragmites. If sea level rise results in increased coastal flooding, native species that cannot survive these conditions may be replaced by more tolerant invasive plants.

B.22.3. THE SOUTH CAROLINA PLAN'S CURRENT INTEGRATION OF CLIMATE CHANGE

Table B-21 summarizes how the South Carolina Aquatic Plant Management Plan addresses and incorporates the projected effects of climate change. South Carolina's Plan does not address climate change in its management and control strategies for aquatic nuisance plants.

Table B-21. Assessment of the South Carolina Aquatic Nuisance Species Management Plan

Aspects of plan that may incorporate climate change	Score
Understanding and incorporating potential impacts resulting from climate change 0 = no; 1 = briefly mentions; 2 = includes general discussion; 3 = includes quantitative info and/or specific examples	
Plan specifically mentions climate change	0
Plan acknowledges climatic boundaries of species	0
Plan demonstrates understanding of species and/or ecosystem sensitivity to changing conditions	0
Plan identifies research on the potential effects of species responding to changing conditions	0
Plan acknowledges regional differences in expected climate changes	0
Capacity to adapt to changing conditions 0 = no; 1 = implicitly (i.e., includes goals and strategies that can be used to account for changing conditions, but does not specify changing conditions as part of their purpose); 2 = yes, explicitly, in passing; 3 = yes, explicitly, and specifies associated goals and/or action items	
Plan accounts for changing conditions in its leadership and coordination goals and strategies	0
Plan accounts for changing conditions in its prevention goals and strategies	0
Plan accounts for changing conditions in its early detection/rapid response goals and strategies	0
Plan accounts for changing conditions in its control and management goals and strategies	0
Plan accounts for changing conditions in its restoration goals and strategies	0
Plan accounts for changing conditions in its research goals and strategies	0
Plan accounts for changing conditions in its information management goals and strategies	0
Plan accounts for changing conditions in its education and public awareness goals and strategies	0
Monitoring strategies 0 = no; 1 = yes, briefly mentions; 2 = yes, but unclear how information will be used; 3 = yes, and specifies associated goals and/or action items	
Plan includes strategy to monitor for changing conditions	0
Plan includes strategy to utilize monitoring data	0
Plan includes strategy for managing/updating monitoring data	1
Revision 0 = no; 1 = yes, in passing; 2 = yes, and includes qualitative description; 3 = yes, and includes timeline and/or benchmarks for doing so	
Plan includes strategy for updating and incorporating new information	2
Funding 0 = no; 1 = a source is specified for a portion of the required funding; 2 = a source is specified for a portion of the required funding along with strategies for obtaining remaining funding; 3 = a source is specified for 100% of required funding	
Plan identifies dedicated funding source for implementation	3
Total score:	**6**

B.22.4. INCORPORATING CLIMATE CHANGE INFORMATION

The goals of South Carolina's Management Plan are limited to aquatic nuisance plants in specific water bodies in the state. The Plan also does not include climate change considerations. Additionally, the Plan lacks prevention, early detection, or monitoring strategies, and relies solely on control measures. Incorporating additional strategies and climate change considerations into current control strategies could improve the Plan's effectiveness.

B.22.4.1. Leadership and Coordination

The Management Plan lacks leadership and coordination strategies to engage its neighbors on the East Coast. Coordination with bordering states, especially those to the south, could help state staff understand what species may be moving north as temperatures rise. Climate change also could make coastal waters warmer, which would affect which AIS can establish and spread. Communicating with neighboring states will allow South Carolina to prepare for these potential new AIS rather than respond once they are already established.

B.22.4.2. Prevention

The Management Plan does not include prevention strategies. Because climate change will alter habitats and affect both natural and human-induced methods of introduction, prevention strategies that incorporate climate change concerns can help to protect the state's water bodies against new AIS introductions.

B.22.4.3. Early Detection/Rapid Response, Control, and Management

Given that climate change will affect sea level rise, coastal erosion, depletion of water quality, and trends in the nutrient cycle in most of South Carolina's aquatic habitats, the Management Plan's control strategies will need to take these changes into account to ensure they remain effective. The Plan should also be expanded to incorporate other types of AIS, and the effect climate change may have on them and their control strategies. Early detection/ rapid response activities should be targeted on new pathways and AIS predicted to become problems under changing conditions.

B.22.4.4. Research

Any research on control strategies should incorporate climate change considerations. Climate change could affect control methods, particularly biocontrol, potentially reducing their effectiveness or causing biocontrol agents themselves to become invasive.

B.23. TEXAS STATE COMPREHENSIVE MANAGEMENT PLAN FOR AQUATIC NUISANCE SPECIES [DRAFT]

B.23.1. GENERAL DESCRIPTION OF TEXAS'S DRAFT PLAN

The draft Texas State Comprehensive Management Plan for Aquatic Nuisance Species (ANS) was under review and awaiting approval by the Governor when this analysis was conducted. The Management Plan was written by the Texas Parks and Wildlife Department (TPWD) (available at http://www.tpwd.state.tx.us/publications/pwdpubs/media/pwd_pl_t3200_1221_draft.doc). The Plan focuses on control and management of ANS that have already been introduced in Texas. The Management Plan is outlines six primary goals to (1) coordinate ANS activities; (2) prevent new introductions; (3) detect, monitor, contain, reduce, or eradicate ANS populations; (4) educate people about preventing introductions and reducing impacts; (5) conduct and disseminate research; and (6) ensure regulations promote ANS prevention and control. Strategic actions and tasks are associated with each goal. A section on implementation lists specific tasks for fiscal years 2006 and 2007.

B.23.2. CLIMATE CHANGE AND AQUATIC INVASIVE SPECIES IN TEXAS

The eastern region of Texas is predicted to experience an estimated increase of 4°F (2.2°C) during the summer and 3.5°F (2°C) in the winter. Precipitation projections are less certain, but general indications are that there might be small changes in winter and substantial increases in summer. Evapotranspiration is projected to increase with rising temperatures, resulting in drier soils and decreased runoff during the growing season (Mulholland et al., 1997).

Increasing water temperatures and decreasing stream flow and groundwater levels may damage native species and may benefit numerous ANS already in the state. For example, the Management Plan notes that hydrilla and water hyacinth have caused great losses to irrigation and drinking water in a variety of areas. Increasing water temperatures will enhance the growth potential of these more tropical ANS. Salt cedar, also, is noted as a problem in the Management Plan: Because salt cedar does well in drier conditions, it may be able to out-compete native plants as conditions become drier.

B.23.3. THE TEXAS DRAFT PLAN'S CURRENT INTEGRATION OF CLIMATE CHANGE

Table B-22 summarizes how the draft Texas State Comprehensive Management Plan for ANS addresses and incorporates the projected effects of climate change. Texas's Plan does not address climate change or changing conditions.

Table B-22. Assessment of the Texas Aquatic Nuisance Species Management Plan

Aspects of plan that may incorporate climate change	Score
Understanding and incorporating potential impacts resulting from climate change 0 = no; 1 = briefly mentions; 2 = includes general discussion; 3 = includes quantitative info and/or specific examples	
Plan specifically mentions climate change	0
Plan acknowledges climatic boundaries of species	0
Plan demonstrates understanding of species and/or ecosystem sensitivity to changing conditions	0
Plan identifies research on the potential effects of species responding to changing conditions	0
Plan acknowledges regional differences in expected climate changes	0
Capacity to adapt to changing conditions 0 = no; 1 = implicitly (i.e., includes goals and strategies that can be used to account for changing conditions, but does not specify changing conditions as part of their purpose); 2 = yes, explicitly, in passing; 3 = yes, explicitly, and specifies associated goals and/or action items	
Plan accounts for changing conditions in its leadership and coordination goals and strategies	0
Plan accounts for changing conditions in its prevention goals and strategies	0
Plan accounts for changing conditions in its early detection/rapid response goals and strategies	0
Plan accounts for changing conditions in its control and management goals and strategies	0
Plan accounts for changing conditions in its restoration goals and strategies	0
Plan accounts for changing conditions in its research goals and strategies	0
Plan accounts for changing conditions in its information management goals and strategies	0
Plan accounts for changing conditions in its education and public awareness goals and strategies	0
Monitoring strategies 0 = no; 1 = yes, briefly mentions; 2 = yes, but unclear how information will be used; 3 = yes, and specifies associated goals and/or action items	
Plan includes strategy to monitor for changing conditions	0
Plan includes strategy to utilize monitoring data	2
Plan includes strategy for managing/updating monitoring data	3
Revision 0 = no; 1 = yes, in passing; 2 = yes, and includes qualitative description; 3 = yes, and includes timeline and/or benchmarks for doing so	
Plan includes strategy for updating and incorporating new information	0
Funding 0 = no; 1 = a source is specified for a portion of the required funding; 2 = a source is specified for a portion of the required funding along with strategies for obtaining remaining funding; 3 = a source is specified for 100% of required funding	
Plan identifies dedicated funding source for implementation	0
Total score:	**5**

B.23.4. INCORPORATING CLIMATE CHANGE INFORMATION

Given Texas's vulnerability to climate change and the impacts invasive species are already having on water availability and flooding, the Management Plan should incorporate climate change concerns. Climate change considerations can be incorporated into several Strategic Actions and Tasks.

B.23.4.1. Leadership and Coordination

Goal 1 calls for coordination of all ANS activities. Climate change should be a consideration in communication on encroaching species and potential vectors and on adapting management practices to accommodate changing conditions.

B.23.4.2. Prevention

Various Tasks under Strategic Action 2A1, which calls for coordination to prevent ANS introductions, should include climate change considerations. For example, Task 2A1b recommends conducting scientific risk assessments to determine priority actions for new ANS threats and to revise earlier assessments. Incorporating climate change information may allow for a more accurate assessment of species' invasibility under changing conditions. Task 2A1d, e, and f call for the TPWD to participate in national conferences; the Gulf of Mexico, Mississippi River, and Western Regional ANS Panels; and the Gulf States Marine Fisheries Commission on ANS issues. Incorporating data from other states and regions into risk analyses will make these analyses more robust and will help track species that may be spreading into Texas.

B.23.4.3. Early Detection/Rapid Response, Control, and Management

Strategic Action 3A1 calls for modifying existing monitoring programs to facilitate early detection. These monitoring programs should incorporate information from risk analyses that account for changes in climate to ensure that threats are adequately monitored.

B.23.4.4. Research

Goal 5 addresses research needs and incorporating climate change considerations into research efforts can inform prioritizations of aquatic invasive species research needs, identifications of habitats vulnerable to future invasions, and improve the development of management strategies that remain effective under changing conditions.

B.24. VIRGINIA INVASIVE SPECIES MANAGEMENT PLAN

B.24.1. GENERAL DESCRIPTION OF VIRGINIA'S PLAN

The Virginia Invasive Species Management Plan includes a significant focus on aquatic invasive species (AIS). In accordance with the Virginia Invasive Species Act of 2003, the Management Plan was written by the Virginia Invasive Species Council in cooperation with the Council's Advisory Committee in 2005 (available at http://www.anstaskforce.gov/State%20Plans/VISMP-final.pdf). The Management Plan outlines seven goals to (1) coordinate efforts; (2) prevent introductions; (3) strengthen and support an early detection network; (4) develop rapid response capabilities; (5) control established invasive species; (6) support or conduct research and risk assessments; and (7) provide information. Under each goal, the Management Plan identifies strategies and needed actions. An implementation table outlines responsible and cooperating agencies, time frames, and costs of planned efforts.

B.24.2. CLIMATE CHANGE AND AQUATIC INVASIVE SPECIES IN VIRGINIA

Temperatures in Virginia are expected to increase by 3 to 4°F (1.6 to 2.2°), depending on the season. With warmer temperatures, there will be less snow and more rain in the winter and more evaporation in the summer, which may lead to lower stream flows and groundwater levels. Algae and eutrophication may become more prevalent as waters warm. Increased precipitation may increase flooding and run-off, which can lower oxygen levels and alter species composition. Sea level rise could also result in habitat loss and salt water intrusion (Moore et al., 1997).

Warmer water temperatures could facilitate the establishment and spread of AIS. For example, the tropical plant hydrilla, found in five counties in central and eastern Virginia, may spread into more western parts of the state as temperatures rise (NRCS, 2007). Other species that are found further south could also migrate to Virginia and successfully overwinter as air and water temperatures increase. Additionally, AIS that have a wider tolerance for environmental conditions, such as the northern snakehead, may out-compete native fish species and become more widespread if water chemistry and temperatures are influenced by climate change.

B.24.3. THE VIRGINIA PLAN'S CURRENT INTEGRATION OF CLIMATE CHANGE

Table B-23 summarizes how the Virginia Invasive Species Management Plan addresses and incorporates the projected effects of climate change. The Management Plan recognizes that climate changes may cause currently non-invasive but non-native species to become invasive. The Plan also recognizes that species adapt to changing conditions. However, the Plan does not address climate change effects on its management actions.

Table B-23. Assessment of the Virginia Aquatic Nuisance Species Management Plan

Aspects of plan that may incorporate climate change	Score
Understanding and incorporating potential impacts resulting from climate change 0 = no; 1 = briefly mentions; 2 = includes general discussion; 3 = includes quantitative info and/or specific examples	
Plan specifically mentions climate change	2
Plan acknowledges climatic boundaries of species	1
Plan demonstrates understanding of species and/or ecosystem sensitivity to changing conditions	1
Plan identifies research on the potential effects of species responding to changing conditions	0
Plan acknowledges regional differences in expected climate changes	0
Capacity to adapt to changing conditions 0 = no; 1 = implicitly (i.e., includes goals and strategies that can be used to account for changing conditions, but does not specify changing conditions as part of their purpose); 2 = yes, explicitly, in passing; 3 = yes, explicitly, and specifies associated goals and/or action items	
Plan accounts for changing conditions in its leadership and coordination goals and strategies	0
Plan accounts for changing conditions in its prevention goals and strategies	0
Plan accounts for changing conditions in its early detection/rapid response goals and strategies	0
Plan accounts for changing conditions in its control and management goals and strategies	0
Plan accounts for changing conditions in its restoration goals and strategies	0
Plan accounts for changing conditions in its research goals and strategies	0
Plan accounts for changing conditions in its information management goals and strategies	0
Plan accounts for changing conditions in its education and public awareness goals and strategies	0
Monitoring strategies 0 = no; 1 = yes, briefly mentions; 2 = yes, but unclear how information will be used; 3 = yes, and specifies associated goals and/or action items	
Plan includes strategy to monitor for changing conditions	0
Plan includes strategy to utilize monitoring data	0
Plan includes strategy for managing/updating monitoring data	0
Revision 0 = no; 1 = yes, in passing; 2 = yes, and includes qualitative description; 3 = yes, and includes timeline and/or benchmarks for doing so	
Plan includes strategy for updating and incorporating new information	3
Funding 0 = no; 1 = a source is specified for a portion of the required funding; 2 = a source is specified for a portion of the required funding along with strategies for obtaining remaining funding; 3 = a source is specified for 100% of required funding	
Plan identifies dedicated funding source for implementation	0
Total score:	7

B.24.4. INCORPORATING CLIMATE CHANGE INFORMATION

Because Virginia will experience various effects from climate change, the Invasive Species Management Plan should incorporate climate change considerations in its next revision, or if the state chooses to develop an aquatic nuisance species (ANS) plan. The current

Management Plan does mention climate change and its potential to affect species' invasiveness. This demonstrates that state agencies may be unaware of the problem. In updating Virginia's Management Plan, the state could further incorporate climate change in a variety of sections, including the following.

B.24.4.1. Leadership and Coordination

Goal 1 calls for coordination of invasive species activities. Climate change should be a consideration in communication on encroaching species and potential vectors and on adapting management practices to accommodate changing conditions.

B.24.4.2. Prevention

Strategy 2.1 identifies pathway analysis and prioritization according to risk. Climate change may affect vectors and these considerations should be incorporated into analyses to improve the effectiveness of prevention activities. Strategic Action 5.1.2 calls for developing restoration plans to establish conditions more suitable for native species. Restoration plans should consider how climate change may affect native habitats and species to ensure that restoration activities are effective. Restoring native habitats may lessen vulnerability to AIS and restoration using species suited to a changing climate may improve resilience over time.

B.24.4.3. Early Detection/Rapid Response, Control, and Management

Strategic Action 3.1.1 calls for surveying and evaluating current monitoring programs and recommending ways to improve detection of invasive species. Identifying how changes in water temperature, water chemistry, and stream flow may influence invasion success will be critical to developing effective monitoring and early detection and rapid response programs.

B.24.4.4. Research

Actions 6.1.1 and 6.1.2 recommend identifying priority research needs for invasive species. Research priorities should include how climate change and invasive species will interact, how this will affect native ecosystems, and how management strategies could be modified to account for the effects of climate change. Working with other states also will help Virginia to identify species that could be moving north. This information should use used to establish a process for assessing potential invasiveness of species that will likely be introduced, as recommended in Action 6.2.4.

B.25. WASHINGTON STATE AQUATIC NUISANCE SPECIES MANAGEMENT PLAN

B.25.1. GENERAL DESCRIPTION OF WASHINGTON'S PLAN

Washington's 2001 Aquatic Nuisance Species (ANS) Management Plan was written by the Washington Department of Fish and Wildlife (WDFW) for the Aquatic Nuisance Species Committee (available at http://wdfw.wa.gov/fish/ans/2001ansplan.pdf) and is a revision to the 1998 Management Plan. The 2001 Plan's goal is to implement a coordinated strategy that minimizes ANS introductions, stops ANS spread, and eradicates or controls ANS to a minimal level of impact. The 2001 Management Plan supports this goal through six objectives to (1) coordinate ANS management; (2) prevent new ANS introductions; (3) detect, monitor, control, or eradicate ANS; (4) educate people about preventing introductions and spread; (5) conduct research on priority species; and (6) adopt regulations that promote prevention and control. Implementation tables are included.

B.25.2. CLIMATE CHANGE AND AQUATIC INVASIVE SPECIES IN WASHINGTON

Average warming in the Pacific Northwest is projected to increase by 3°F (1.7°C) by the 2020s and 5°F (2.8°C) by the 2050s. Annual precipitation projections are less certain, ranging from 7% or 2 inches (5cm) to a 13% or 4 inches (10cm). Precipitation increases will occur in winter, with decreases (or smaller increases) in summer; for this reason, projections that show increases in annual precipitation also show decreases in water availability (Parson, 2001b). Sea level rise projects are greater for the Pacific than the Atlantic coast. Higher mean sea level may increase sediment erosion and redistribution on the open coast (Parson, 2001b).

Climate change will cause continued changes in coastal and estuarine ecosystems through changes in runoff and increased water temperatures, potentially increasing ANS introductions. The warm decade of the 1980s, following the shift to warm-phase Pacific Decadal Oscillation in the late 1970s, was characterized by a rapid expansion of exotic cordgrass (Spartina) in Willapa Bay that threatened local species (Parson, 2001b). Species tolerant of a wide range of water salinity and temperatures such as the Asian crab could become a greater threat in estuaries and tidal areas as salt water intrudes into freshwater areas with sea level rise.

B.25.3. THE WASHINGTON PLAN'S CURRENT INTEGRATION OF CLIMATE CHANGE

Table B-24 summarizes how Washington's ANS Management Plan addresses and incorporates the projected effects of climate change. Although the Management Plan does not explicitly mention climate change, it does recognize that certain species have climatic boundaries. Additionally, it offers extensive strategies for preventing and monitoring ANS.

Table B-24. Assessment of the Washington Aquatic Nuisance Species Management Plan

Aspects of plan that may incorporate climate change	Score
Understanding and incorporating potential impacts resulting from climate change 0 = no; 1 = briefly mentions; 2 = includes general discussion; 3 = includes quantitative info and/or specific examples	
Plan specifically mentions climate change	0
Plan acknowledges climatic boundaries of species	3
Plan demonstrates understanding of species and/or ecosystem sensitivity to changing conditions	0
Plan identifies research on the potential effects of species responding to changing conditions	0
Plan acknowledges regional differences in expected climate changes	0
Capacity to adapt to changing conditions 0 = no; 1 = implicitly (i.e., includes goals and strategies that can be used to account for changing conditions, but does not specify changing conditions as part of their purpose); 2 = yes, explicitly, in passing; 3 = yes, explicitly, and specifies associated goals and/or action items	
Plan accounts for changing conditions in its leadership and coordination goals and strategies	0
Plan accounts for changing conditions in its prevention goals and strategies	0
Plan accounts for changing conditions in its early detection/rapid response goals and strategies	0
Plan accounts for changing conditions in its control and management goals and strategies	0
Plan accounts for changing conditions in its restoration goals and strategies	0
Plan accounts for changing conditions in its research goals and strategies	2
Plan accounts for changing conditions in its information management goals and strategies	0
Plan accounts for changing conditions in its education and public awareness goals and strategies	1
Monitoring strategies 0 = no; 1 = yes, briefly mentions; 2 = yes, but unclear how information will be used; 3 = yes, and specifies associated goals and/or action items	
Plan includes strategy to monitor for changing conditions	0
Plan includes strategy to utilize monitoring data	3
Plan includes strategy for managing/updating monitoring data	3
Revision 0 = no; 1 = yes, in passing; 2 = yes, and includes qualitative description; 3 = yes, and includes timeline and/or benchmarks for doing so	
Plan includes strategy for updating and incorporating new information	3
Funding 0 = no; 1 = a source is specified for a portion of the required funding; 2 = a source is specified for a portion of the required funding along with strategies for obtaining remaining funding; 3 = a source is specified for 100% of required funding	
Plan identifies dedicated funding source for implementation	2
Total score:	**17**

B.25.4. INCORPORATING CLIMATE CHANGE INFORMATION

Given the projected effects of climate change in Washington, the state's Management Plan should incorporate climate change-related actions and strategies. State staff could incorporate climate change-related data, criteria, and models in implementing the existing tasks outlined in the Management Plan.

B.25.4.1. Leadership and Coordination

Objective 1 calls for coordination of invasive species activities. Climate change should be a consideration in communication on encroaching species and potential vectors and on adapting management practices to accommodate changing conditions.

B.25.4.2. Prevention

Strategic Action 2A3 calls for prohibiting, controlling, and permitting non-native aquatic species based on their invasive potential. Task 2A3a calls for the development of a screening process to determine invasive potential. This process should incorporate climate change considerations. For example, an ANS that cannot currently overwinter in the state may persist if its water temperature tolerance range overlaps with projected temperature increases.

B.25.4.3. Early Detection/Rapid Response, Control, and Management

Priorities for managing existing ANS (Strategic Action 3C2) should take climate change effects into consideration. Species currently in Washington may spread and become a greater problem with warmer water temperatures or salt water intrusion.

B.25.4.4. Research

Task 2A3d calls for the WDFW and the ANS Coordinating Committee to develop and implement a process to identify threats to state waters, including threats from the spread of existing ANS, and to assess the environmental risks associated with each threat. The model should incorporate how climate change impacts may exacerbate threats from ANS. For example, water hyacinth could be better suited to survive in Washington if water temperatures become warmer due to climate change. Impacts from water hyacinth could also compound impacts from climate change. If climate change results in warmer summer temperatures and droughts, then water availability could be reduced. Water hyacinth could further reduce water supply for both domestic and industry uses by blocking intake pipes.

B.26. WISCONSIN COMPREHENSIVE MANAGEMENT PLAN TO PREVENT FURTHER INTRODUCTIONS AND CONTROL EXISTING POPULATIONS OF AQUATIC INVASIVE SPECIES

B.26.1. GENERAL DESCRIPTION OF WISCONSIN'S PLAN

The Wisconsin Department of Natural Resources, the University of Wisconsin Sea Grant program, and the Great Lakes Indian Fish and Wildlife Commission partnered to develop the Wisconsin Comprehensive Management Plan, which was released in 2003 (available at http://dnr.wi.gov/invasives/compstateansplanfinal0903.pdf). The major goals of the Wisconsin Plan are Aquatic Nuisance Species (ANS) prevention, control, and abatement. As of 2003, the Great Lakes contained 163 exotic species of fish, invertebrates, pathogens, and plant species. The Management Plan outlines strategies to combat existing ANS as well as to protect inland waters from similar threats. The Management Plan also acts as a funding proposal to the national ANS Task Force.

B.26.2. CLIMATE CHANGE AND AQUATIC INVASIVE SPECIES IN WISCONSIN

Temperatures are projected to increase by 5 to 10°F (2.8 to 5.5°C) in the Midwest region by 2100. Precipitation is expected to increase by approximately 10 to 30%. Increasing temperatures are expected to increase evaporation, triggering a soil moisture deficit, reduction in lake and river water levels, and diminished groundwater recharge. For smaller lakes and rivers, reduced flows are likely to intensify water quality issues. In particular, eutrophication of lakes will likely increase due to increases in excess nutrient runoff from heavy precipitation events and warmer lake temperatures that stimulate algae growth (Easterling and Karl, 2001).

As water temperatures in lakes increase, significant changes in freshwater ecosystems may occur. The zoogeographical boundary for fish species may shift north by 500 to 600 kilometers, leading to invasions by warmer water fishes and extirpations of colder water fishes (Magnuson et al., 1997; Easterling and Karl, 2001). Temperature may limit the extent of zebra mussel colonization and has thus far kept populations in Lake Superior small. During the breeding season, when the water temperature is above 54°F (12°C), each mature female can produce several hundred thousand eggs. The longer this period, the more successful colonization will be. With summer water temperatures increasing in northern lakes, the currently small colonies may become more widespread (Easterling and Karl, 2001).

B.26.3. THE WISCONSIN PLAN'S CURRENT INTEGRATION OF CLIMATE CHANGE

Table B-25 summarizes how the Wisconsin Comprehensive Management Plan addresses and incorporates the projected effects of climate change. Although the Management Plan does not explicitly address climate change, it does recognize monitoring and regularly updating information on local and surrounding conditions as integral to successful ANS management.

Table B-25. Assessment of the Wisconsin Aquatic Nuisance Species Management Plan

Aspects of plan that may incorporate climate change	Score
Understanding and incorporating potential impacts resulting from climate change 0 = no; 1 = briefly mentions; 2 = includes general discussion; 3 = includes quantitative info and/or specific examples	
Plan specifically mentions climate change	0
Plan acknowledges climatic boundaries of species	1
Plan demonstrates understanding of species and/or ecosystem sensitivity to changing conditions	0
Plan identifies research on the potential effects of species responding to changing conditions	0
Plan acknowledges regional differences in expected climate changes	0
Capacity to adapt to changing conditions 0 = no; 1 = implicitly (i.e., includes goals and strategies that can be used to account for changing conditions, but does not specify changing conditions as part of their purpose); 2 = yes, explicitly, in passing; 3 = yes, explicitly, and specifies associated goals and/or action items	
Plan accounts for changing conditions in its leadership and coordination goals and strategies	0
Plan accounts for changing conditions in its prevention goals and strategies	0
Plan accounts for changing conditions in its early detection/rapid response goals and strategies	0
Plan accounts for changing conditions in its control and management goals and strategies	0
Plan accounts for changing conditions in its restoration goals and strategies	0
Plan accounts for changing conditions in its research goals and strategies	0
Plan accounts for changing conditions in its information management goals and strategies	0
Plan accounts for changing conditions in its education and public awareness goals and strategies	0
Monitoring strategies 0 = no; 1 = yes, briefly mentions; 2 = yes, but unclear how information will be used; 3 = yes, and specifies associated goals and/or action items	
Plan includes strategy to monitor for changing conditions	0
Plan includes strategy to utilize monitoring data	3
Plan includes strategy for managing/updating monitoring data	0
Revision 0 = no; 1 = yes, in passing; 2 = yes, and includes qualitative description; 3 = yes, and includes timeline and/or benchmarks for doing so	
Plan includes strategy for updating and incorporating new information	1
Funding 0 = no; 1 = a source is specified for a portion of the required funding; 2 = a source is specified for a portion of the required funding along with strategies for obtaining remaining funding; 3 = a source is specified for 100% of required funding	
Plan identifies dedicated funding source for implementation	3
Total score:	**8**

B.26.4. INCORPORATING CLIMATE CHANGE INFORMATION

Given the projected effects of climate change for Wisconsin, the Management Plan should incorporate climate change considerations when revising the Plan to improve prevention and management strategies.

B.26.4.1. Leadership and Coordination

The Management Plan describes that partnerships with interstate and international groups will be formed to promote consistent regional approaches to aquatic invasive species (AIS) management. Strategy II E also calls for working with partners to improve coordination efforts on AIS. Because a species' ability to spread is affected in part by climate, state staff should consider changes in water and air temperatures when assessing invasion threats and developing management strategies.

B.26.4.2. Prevention

Goal I calls for the implementation of procedures and practices to prevent new introductions of AIS into Lakes Michigan and Superior, the state's boundary waters, and inland water systems. The Management Plan focuses its prevention strategies on primary vectors including the sale and distribution of bait, the aquaculture and aquarium industries, and ballast water discharges. Warming waters, altered hydrology, and nutrient level changes may affect the ability of certain aquarium species or bait fish to survive and become established in Wisconsin's waters. Incorporating these climate change considerations into prevention strategies may improve their success.

B.26.4.3. Early Detection/Rapid Response, Control, and Management

Goal II of the Management Plan calls for implementing of management strategies that limit the spread of established AIS into uninfested waters. State staff should consider how control strategies may be affected by climate change and adjust them accordingly. For example, current control methods for an AIS restricted to a small part of the state may be effective in preventing its spread. However, as water temperatures warm that AIS' invasive potential could increase. Furthermore, monitoring should incorporate climate change considerations to ensure that the appropriate species and introduction points are targeted during early detection activities.

B.26.4.4. Research

The Management Plan does not specifically address research needs, except for ballast water technologies. Any research on the invasion pathway and potential AIS should also consider the implications of climate change on the pathway and species.

REFERENCES

Christensen, JH; Hewitson, B; Busuioc, A; et al. (2007) Regional climate projections. In: Solomon, S; Qin, D; Manning, M; et al., eds. Climate change 2007: the physical science basis. Contribution of Working Group I to the Fourth Assessment Report of the Intergovernmental Panel on Climate Change. Cambridge, United Kingdom: Cambridge University Press; pp. 847–940.

Covich, AP; Fritz, SC; Lamb, PJ; et al.(1997) Potential effects of climate change on aquatic ecosystems of the Great Plains of North America. Hydrol Processes 11(8):993–1021

Easterling, DR; Karl, TR. (2001) Potential consequences of climate variability and change for the midwestern United States. In: Climate change impacts on the United States: the potential consequences of climate variability and change. Report for the US Global Change Research Program. National Assessment Synthesis Team. Cambridge, United Kingdom: Cambridge University Press; pp. 167–188. Available online at http://www.gcrio.org/NationalAssessment/6MW.pdf [accessed May 1, 2007].

Harvell, CD; Kim, K; Burkholder, JM; et al. (1999) Emerging marine diseases-climate links and anthropogenic factors. Science 285(5433):1505–1510.

Harvell, CD; Mitchell, CE; Ward, JR; et al. (2002) Climate warming and disease risks for terrestrial and marine biota. Science 296(5576):2158–2162.

Hayhoe, K; Wake, CP; Huntington, TG; et al. (2007) Past and future changes in climate and hydological indicators in the US Northeast. Clim Dyn 28:381–407.

Jones, RJ; Bowyer, J; Hoegh-Guldberg, O; et al. (2004) Dynamics of a temperature-related coral disease outbreak. Mar Ecol Prog Ser 281:63–77.

Joyce, LA; Ojima, D; Seielstaf, GA; et al. (2001) Potential consequences of climate variability and change for the Great Plains. In: Climate change impacts on the United States: the potential consequences of climate variability and change. Report for the US Global Change Research Program. National Assessment Synthesis Team. Cambridge, United Kingdom: Cambridge University Press; pp. 191–217. Available online at http://www.usgcrp.gov/usgcrp/Library/nationalassessment/07GP.pdf [accessed May 1, 2007].

KDHE (Kansas Department of Health and Environment). Zebra mussel: their inevitable arrival in Kansas. Available online at http://www.kdheks.gov/befs/download/zebra_mussel_article.pdf [accessed May 1, 2007].

LaCoast. (2003) Grappling with the unknown: how much of a change do climate models project? Funded by the US Coastal Planning, Protection and Restoration Act (CWPPRA). Available online at http http://lacoast.gov/watermarks/2003-02/2howmuch/ [accessed May 1, 2007].

Magnuson, JJ; Webster, KE; Assel, RA; et al. (1997) Potential effects of climate changes on aquatic systems: Laurentian Great Lakes and Precambrian Shield Region. Hydrol Processes 11(8):825–871.

Moore, MV; Pace, ML; Mather, JR; et al. (1997) Potential effects of climate change on freshwater ecosystems of the New England/Mid-Atlantic Region. Hydrol Processes 11(8):925–947.

Moser, S; Hayhoe, K; Wander, M. (2004) Climate change in the Hawkeye State: impacts on Iowa communities and ecosystems. Union of Concerned Scientists; pp.16. Available online at http://www.ucsusa.org/assets/documents/clean_energy/Climate_Change_in_Iowa_-Long-_Final-_and_Formatted.pdf.

Mulholland, PJ; Best, GR; Coutant, CC; et al. (1997) Effects of climate change on freshwater ecosystems of the south-eastern United States and the Gulf Coast of Mexico. Hydrol Processes 11(8):949–970.

NPWRC (Northern Prairie Wildlife Research Center). (2006) Species abstracts of highly disruptive exotic plants. US Geological Survey. Available online at http://www.npwrc.usgs.gov/resource/plants/exoticab/index.htm [accessed May 1, 2007].

NRCS (Natural Resources Conservation Service). Plants profile. US Department of Agriculture. Available online at http://plants.usda.gov/java/county?state_name=Virginia&statefips=51&symbol=HYVE3 [accessed May 1, 2007].

Parson, EA. (2001a) Potential consequences of climate variability and change for Alaska. In: Climate change impacts on the United States: the potential consequences of climate variability and change. Report for the US Global Change Research Program. National Assessment Synthesis Team. Cambridge, United Kingdom: Cambridge University Press; pp. 283–312. Available online at http://www.usgcrp.gov/usgcrp/Library/nationalassessment/10Alaska.pdf [accessed May 1, 2007].

Parson, EA. (2001b) Potential consequences of climate variability and change for the Pacific Northwest. In: Climate change impacts on the United States: the potential consequences of climate variability and change. Report for the US Global Change Research Program. National Assessment Synthesis Team. Cambridge, United Kingdom: Cambridge University Press; pp. 247–280. Available online at http://www.usgcrp.gov/usgcrp/Library/nationalassessment/09PNW.pdf [accessed May 1, 2007].

PIRAG (Pacific Island Regional Assessment Group). (2001) Pacific Islands Region. In: Preparing for a changing climate: the potential consequences of climate variability and change. US Global Change Research Program; pp. 2–24. Available online at http://www2.eastwestcenter.org/climate/assessment/ch2a.pdf [accessed May 2, 2007].

SRAG (Southwest Regional Assessment Group). (2000) Preparing for a changing climate: the potential consequences of climate variability and change. Southwest. US Global Change Research Program. Available online at http://www.ispe.arizona.edu/research/swassess/report.html [accessed Apr. 30, 2007].

Seager, R; Ting, M; Held, I; et al. (2007) Model projections of an imminent transition to a more arid climate in southwestern North America. Science 316(5828):1181–1184.

USACE (United.States Army Corps of Engineers). Round Goby – *Neogobius melanostomus*. Available online at http://el.erdc.usace.army.mil/ansrp/neogobius_melanostomus.pdf [accessed May 1, 2007].

APPENDIX C

REGIONAL AQUATIC INVASIVE SPECIES MANAGEMENT PLAN SUMMARIES

CONTENTS

LIST OF TABLES

C.1. METHODS

We reviewed regional aquatic invasive species (AIS) management plans, where available, and assessed how regional AIS panels address climate change specifically, as well as how they generally provide for adaptation of strategies and actions under changing conditions. There are a total of seven regional plans. Several regional plans refer to AIS as aquatic nuisance species (ANS). To maintain consistency with regional plan language, this appendix generally uses ANS as a synonym for AIS. Regional plans examined include the following:

- Great Lakes Action Plan for the Prevention and Control of Non-indigenous Aquatic Nuisance Species

- Gulf of Mexico Aquatic Nuisance Species in the Gulf of Mexico: A Guide for Future Action by the Gulf of Mexico Regional Panel and the Gulf States

- Lake Champlain Basin Aquatic Nuisance Species Management Plan

- Midwest Region Aquatic Nuisance Species Action Plan

- Northeast Region Aquatic Nuisance Species Action Plan

- Southeast Region Aquatic Nuisance Species Action Plan

- Western Region Aquatic Nuisance Species Action Plan

The following summaries also include ways to incorporate climate considerations and adaptive management procedures into individual plan goals and strategies when revising the action plans.

C.2. GREAT LAKES ACTION PLAN FOR THE PREVENTION AND CONTROL OF NON-INDIGENOUS AQUATIC NUISANCE SPECIES

C.2.1. GENERAL DESCRIPTION OF THE GREAT LAKES PLAN

The Great Lakes Action Plan was written by the Great Lakes Panel on Aquatic Nuisance Species (ANS) (available at http://glc.org/ans/pubs.html). The goals of the plan are to prevent ANS introductions, limit ANS spread, and minimize impacts. In an addendum, the Action Plan lists objectives and strategic actions designed to achieve these goals through management; research and monitoring; and information, education, and collaboration.

C.2.2. CLIMATE CHANGE AND INVASIVE SPECIES IN THE GREAT LAKES

Climate models project temperatures increases of 5 to10°F (2.8 to 5.5°C) and precipitation increases of 10 to 30%in the Midwest region by 2100 (Easterling and Karl, 2001). Increasing temperatures are expected to increase evaporation, triggering a soil moisture deficit, reduction in lake and river water levels, and diminished groundwater recharge. Models project lake level losses of up to 5 feet (1.5 meters) in the Great Lakes. In smaller lakes and rivers, increased evaporation will cause reduced flows, exacerbating water quality issues (Magnuson et al., 1997; Easterling and Karl, 2001). Heavy precipitation events are also expected to increase, washing nutrients and runoff from urban, agriculture, and construction sites into waterways (Magnuson et al., 1997).

Warmer water temperatures may make waterways more vulnerable to ANS invasions; aquatic ecosystems in the Midwest may experience a shift from coldwater fish species such as trout to warmer water species such as bass and catfish (Robillard and Fox, 2006; Jacobs et al., 2001). For example, the Lake Superior Lakewide Management Plan notes that environmental conditions in the Lake Superior basin have prevented reproduction of zebra mussels to date, but that mild weather in recent years has allowed reproduction to occur in the St. Louis estuary. As the climate in the region warms, conditions in the Lake Superior basin may be more amenable to zebra mussel reproduction, allowing the species to become established in the ecosystem (U.S. EPA, 2000).

C.2.3. THE GREAT LAKE PLAN'S CURRENT INTEGRATION OF CLIMATE CHANGE

Table C-1 summarizes how the Great Lakes Action Plan for the Prevention and Control of Nonindigenous Aquatic Nuisance Species addresses and incorporates the projected effects of climate change. The Great Lakes Plan does not address the effects of climate change or changing environmental conditions in general in its recommendations, objectives, and/or strategic actions.

Table C-1. Assessment of the Great Lakes Aquatic Nuisance Species Action Plan

Aspects of plan that may incorporate climate change	Score
Understanding and incorporating potential impacts resulting from climate change 0 = no; 1 = briefly mentions; 2 = includes general discussion; 3 = includes quantitative info and/or specific examples	
Plan specifically mentions climate change	0
Plan acknowledges climatic boundaries of species	0
Plan demonstrates understanding of species and/or ecosystem sensitivity to changing conditions	0
Plan identifies research on the potential effects of species responding to changing conditions	0
Plan acknowledges regional differences in expected climate changes	0
Capacity to adapt to changing conditions 0 = no; 1 = implicitly (i.e., includes goals and strategies that can be used to account for changing conditions, but does not specify changing conditions as part of their purpose); 2 = yes, explicitly, in passing; 3 = yes, explicitly, and specifies associated goals and/or action items	
Plan accounts for changing conditions in its leadership and coordination goals and strategies	0
Plan accounts for changing conditions in its prevention goals and strategies	0
Plan accounts for changing conditions in its early detection and rapid response goals and strategies	0
Plan accounts for changing conditions in its control and management goals and strategies	0
Plan accounts for changing conditions in its restoration goals and strategies	0
Plan accounts for changing conditions in its research goals and strategies	0
Plan accounts for changing conditions in its information management goals and strategies	0
Plan accounts for changing conditions in its education and public awareness goals and strategies	0
Monitoring strategies 0 = no; 1 = yes, briefly mentions; 2 = yes, but unclear how information will be used; 3 = yes, and specifies associated goals and/or action items	
Plan includes strategy to monitor for changing conditions	0
Plan includes strategy to utilize monitoring data	0
Plan includes strategy for managing/updating monitoring data	0
Revision 0 = no; 1 = yes, in passing; 2 = yes, and includes qualitative description; 3 = yes, and includes timeline and/or benchmarks for doing so	
Plan includes strategy for updating and incorporating new information	0
Funding 0 = no; 1 = a source is specified for a portion of the required funding; 2 = a source is specified for a portion of the required funding along with strategies for obtaining remaining funding; 3 = a source is specified for 100% of required funding	
Plan identifies dedicated funding source for implementation	0
Total score:	**0**

C.2.4. INCORPORATING CLIMATE CHANGE INFORMATION

Climate change will affect the Great Lakes region and its water resources throughout the next decades. The ANS Action Plan should incorporate climate change effects so that management activities can better respond to changing conditions and remain effective over time. Several of the Strategic Actions listed in the Action Plan could incorporate climate change considerations to adapt management activities in the Great Lakes region.

C.2.4.1. Leadership and Coordination

Collaboration objectives call for inter-jurisdictional cooperation on prevention and control measures, regulation, and education efforts. Coordination on the regional level offers the opportunity for states to share information on climate-related ANS issues, including expanding ANS ranges, changing conditions, and adapting management to reflect changing conditions. These measures will not only help states collaboratively identify species of concern, but incorporation of these measures also better prevents the establishment and spread of ANS.

C.2.4.2. Prevention

Management objectives call for the assessment and characterization of pathways for ANS and the identification of high risk ANS. Efforts to assess risk and develop species lists should account for the effect of climate change on ecosystem vulnerability to ANS invasion and expanding species ranges and vectors.

C.2.4.3. Early Detection and Rapid Response, Control, and Management

Several Strategic actions address ANS control. Climate change considerations should be included in developing control strategies for the Great Lakes Region that remain effective over time.

C.2.4.4. Research

Research objectives also call for the development of management strategies that address ANS threats. However, if ecosystem conditions change over time, management practices may lose effectiveness. Research should address changing conditions in order to better inform management practices to ensure that they remain robust in the context of a changing climate.

C.3. A GUIDE FOR FUTURE ACTION BY THE GULF OF MEXICO REGIONAL PANEL AND THE GULF STATES

C.3.1. GENERAL DESCRIPTION OF THE GULF OF MEXICO'S PLAN

The Gulf of Mexico's 2003 Guide for Future Action examines the structure and activities of the Gulf of Mexico Regional Panel (GMRP) and was developed by the Sea Grant Law Center (available at http://www.olemiss.edu/orgs/SGLC/ANS.pdf). The Guide summarizes current aquatic nuisance species (ANS) actions and offers recommendations for ANS management for the individual Gulf States. It should be noted that the Regional Panel did not develop the Guide and does *not* use the Guide as guidance. Actions are divided into the categories of coordination, prevention, regulation, control and management, and enforcement and implementation. The Guide outlines recommended actions for each state to better address ANS threats and for the GMRP to better support interstate cooperation and the development and implementation of regional priorities.

C.3.2. CLIMATE CHANGE AND INVASIVE SPECIES IN THE GULF OF MEXICO

Climate change projections for the Southeastern United States, including some Gulf of Mexico states, vary; although most climate models project warming temperatures (Mearns et al., 2003; NAST, 2001). Climate change models also project a 10 to 25% increase in precipitation (Burkett et al., 2001; Mulholland et al., 1997). Increased temperatures will likely increase stress on water quality. Waterbodies in the Southeast already receive pollution from agriculture, urban areas, and mining. Increased precipitation and more frequent, extreme precipitation events will flush more contaminated run-off into waterbodies, and higher temperatures will reduce dissolved oxygen levels in water (Gibson et al., 2005; Jacobs et al., 2001).

Increased precipitation will also likely lead to higher freshwater inflows into estuaries and lower salinity, although sea level rise could increase salinity levels (Mulholland et al., 1997). The region's salt marshes will be particularly affected by sea level rise and other climate change factors, leading to changes in salinity and nutrient availability (Burkett and Kusler, 2000). These ecosystem changes could provide more suitable conditions for ANS establishment and spread.

C.3.3. THE GULF OF MEXICO PLAN'S CURRENT INTEGRATION OF CLIMATE CHANGE

Table C-2 summarizes how Gulf of Mexico's Guide for Future Action addresses and incorporates the projected effects of climate change. The recommendations in the Guide are state-specific and generally focused on increasing state agency jurisdiction and resources, as opposed to specific management tasks. The Guide does not explicitly address climate change, but it does support the use of monitoring programs to identify and track invasions, and therefore offers opportunities to consider changing conditions.

Table C-2. Assessment of the Gulf of Mexico Aquatic Nuisance Species Guide for Future Action

Aspects of plan that may incorporate climate change	Score
Understanding and incorporating potential impacts resulting from climate change 0 = no; 1 = briefly mentions; 2 = includes general discussion; 3 = includes quantitative info and/or specific examples	
Plan specifically mentions climate change	0
Plan acknowledges climatic boundaries of species	0
Plan demonstrates understanding of species and/or ecosystem sensitivity to changing conditions	0
Plan identifies research on the potential effects of species responding to changing conditions	0
Plan acknowledges regional differences in expected climate changes	0
Capacity to adapt to changing conditions 0 = no; 1 = implicitly (i.e., includes goals and strategies that can be used to account for changing conditions, but does not specify changing conditions as part of their purpose); 2 = yes, explicitly, in passing; 3 = yes, explicitly, and specifies associated goals and/or action items	
Plan accounts for changing conditions in its leadership and coordination goals and strategies	0
Plan accounts for changing conditions in its prevention goals and strategies	0
Plan accounts for changing conditions in its early detection and rapid response goals and strategies	0
Plan accounts for changing conditions in its control and management goals and strategies	0
Plan accounts for changing conditions in its restoration goals and strategies	0
Plan accounts for changing conditions in its research goals and strategies	0
Plan accounts for changing conditions in its information management goals and strategies	0
Plan accounts for changing conditions in its education and public awareness goals and strategies	0
Monitoring strategies 0 = no; 1 = yes, briefly mentions; 2 = yes, but unclear how information will be used; 3 = yes, and specifies associated goals and/or action items	
Plan includes strategy to monitor for changing conditions	0
Plan includes strategy to utilize monitoring data	1
Plan includes strategy for managing/updating monitoring data	0
Revision 0 = no; 1 = yes, in passing; 2 = yes, and includes qualitative description; 3 = yes, and includes timeline and/or benchmarks for doing so	
Plan includes strategy for updating and incorporating new information	0
Funding 0 = no; 1 = a source is specified for a portion of the required funding; 2 = a source is specified for a portion of the required funding along with strategies for obtaining remaining funding; 3 = a source is specified for 100% of required funding	
Plan identifies dedicated funding source for implementation	0
Total score:	**1**

C.3.4. INCORPORATING CLIMATE CHANGE INFORMATION

In light of the considerable impacts of climate change expected in the Gulf of Mexico region, especially the effects of rising sea levels on coastal marshes, the GMRP should address climate change in its future efforts to coordinate state management activities. Several GMRP actions listed in the Guide could incorporate climate change considerations to adapt management activities in the Gulf of Mexico Region.

C.3.4.1. Leadership and Coordination

The Guide recommends that the GMRP issue policy statements to identify regional goals and priorities, focus and direct state goals, and urge states to direct resources toward certain activities in order to facilitate regional coordination. These regional goals and priorities should address climate change. Sharing effective management strategies may be very helpful for other states experiencing similar impacts of changing conditions, such as sea level rise and effects on regionally spread ANS such as salvinia and nutria. The Guide also encourages states to develop comprehensive ANS management plans. The development of new plans offers the opportunity incorporate the effects of climate change. Risk assessments of ANS threats based on the effects of climate change and habitat restoration in climate change-impacted salt marshes, for example, could be incorporated into these plans.

C.3.4.2. Prevention

The Guide also addresses states' assessments of ANS risks. The Guide recommends that Florida establish a program for risk assessment, and that Alabama, Mississippi, Louisiana, and Texas establish "clean lists" of species allowed into the state. Assessment of the invasion risk posed by a non-native species should take into account changing conditions that may.

C.3.4.3. Early Detection and Rapid Response, Control, and Management

The Guide suggests that the GMRP should develop model plans for early detection and rapid response for states to incorporate into their individual plans. These model plans provide an opportunity to incorporate climate change considerations that can inform states on detecting new ANS that might invade as environmental conditions change.

C.3.4.4. Research

The Guide describes on-going activities of the GMRP, including working group tasks. These tasks include identifying research needs. Working groups should address climate change effects as research needs that state ANS management plans should incorporate.

C.4. LAKE CHAMPLAIN BASIN AQUATIC NUISANCE SPECIES MANAGEMENT PLAN

C.4.1. GENERAL DESCRIPTION OF LAKE CHAMPLAIN'S PLAN

The Lake Champlain Basin Aquatic Nuisance Species (ANS) Management Plan was approved in May 2000 (available at http://www.northeastans.org/pdf/lcbansplan2000.PDF). The Management Plan was coordinated by the Vermont Department of Environmental Conservation and the New York State Department of Environmental Conservation in cooperation with state and federal agencies, lake groups, and the research communities in New York, Vermont, and Quebec. The three goals of the plan are to prevent new introductions, limit ANS spread into uninfested waters, and abate ANS impacts. In order to accomplish these goals, the Management Plan outlines six objectives to (1) coordinate plan implementation; (2) fill information gaps; (3) select target ANS; (4) evaluate ANS management alternatives; (5) implement management actions that eradicate and/or prevent the spread of ANS; and (6) increase awareness ANS issues. Each objective has associated strategies and actions. The Management Plan also describes ANS problems in the Lake Champlain Basin, authorities and programs related to ANS management, and a list of priority actions for implementation in the two years following the release of the Plan.

C.4.2. CHANGE AND INVASIVE SPECIES IN LAKE CHAMPLAIN

Climate change models project that minimum winter temperatures in the Northeast may increase from 5 to 9.5°F (2.9 to 5.3°C) by 2100 (Hayhoe et al., 2007). Precipitation projections range from increases of up to 30% in the summer to decreases or small changes in the winter (Hayhoe et al., 2007). These climatic changes are likely to increase water temperature, which will affect dissolved oxygen levels and nutrient content (Hayhoe et al., 2007).

The precipitation regime greatly influences the lake's ecosystem and changes to this regime d could negatively affect habitat for lake species and facilitate the spread of ANS already present in the Lake Champlain Basin. For example, the common reed (*Phragmites australis*) is identified in the Management Plan as an ANS already established in the basin. Altered conditions and habitat availability due to climate change may allow Phragmites to spread further.

C.4.3. THE LAKE CHAMPLAIN PLAN'S CURRENT INTEGRATION OF CLIMATE CHANGE

Table C-3 summarizes how the Lake Champlain Basin ANS Management Plan addresses and incorporates the projected effects of climate change. While the Management Plan does not address climate change effects, it does include a monitoring strategy for ANS. Furthermore, the Plan's framework does present opportunities to introduce climate change considerations.

Table C-3. Assessment of the Lake Champlain Basin Aquatic Nuisance Species Management Plan

Aspects of plan that may incorporate climate change	Score
Understanding and incorporating potential impacts resulting from climate change 0 = no; 1 = briefly mentions; 2 = includes general discussion; 3 = includes quantitative info and/or specific examples	
Plan specifically mentions climate change	0
Plan acknowledges climatic boundaries of species	0
Plan demonstrates understanding of species and/or ecosystem sensitivity to changing conditions	0
Plan identifies research on the potential effects of species responding to changing conditions	0
Plan acknowledges regional differences in expected climate changes	0
Capacity to adapt to changing conditions 0 = no; 1 = implicitly (i.e., includes goals and strategies that can be used to account for changing conditions, but does not specify changing conditions as part of their purpose); 2 = yes, explicitly, in passing; 3 = yes, explicitly, and specifies associated goals and/or action items	
Plan accounts for changing conditions in its leadership and coordination goals and strategies	0
Plan accounts for changing conditions in its prevention goals and strategies	0
Plan accounts for changing conditions in its early detection and rapid response goals and strategies	1
Plan accounts for changing conditions in its control and management goals and strategies	1
Plan accounts for changing conditions in its restoration goals and strategies	0
Plan accounts for changing conditions in its research goals and strategies	0
Plan accounts for changing conditions in its information management goals and strategies	0
Plan accounts for changing conditions in its education and public awareness goals and strategies	0
Monitoring strategies 0 = no; 1 = yes, briefly mentions; 2 = yes, but unclear how information will be used; 3 = yes, and specifies associated goals and/or action items	
Plan includes strategy to monitor for changing conditions	0
Plan includes strategy to utilize monitoring data	3
Plan includes strategy for managing/updating monitoring data	3
Revision 0 = no; 1 = yes, in passing; 2 = yes, and includes qualitative description; 3 = yes, and includes timeline and/or benchmarks for doing so	
Plan includes strategy for updating and incorporating new information	3
Funding 0 = no; 1 = a source is specified for a portion of the required funding; 2 = a source is specified for a portion of the required funding along with strategies for obtaining remaining funding; 3 = a source is specified for 100% of required funding	
Plan identifies dedicated funding source for implementation	1
Total score:	**12**

C.4.4. INCORPORATING CLIMATE CHANGE INFORMATION

Climate change will affect the Lake Champlain Basin and its water resources throughout the next decades. The ANS Action Plan should incorporate climate change effects so that management activities can better respond to changing conditions and remain effective over time. Several of the Strategic Objectives listed in the Management Plan could incorporate climate change considerations to adapt management activities in the Lake Champlain Basin.

C.4.4.1. Leadership and Coordination

Objective A discusses coordination of plan implementation. Strategies under this Objective should include measures to coordinate among states on new ANS and range expansions that may be due to climate change.

C.4.4.2. Prevention

Objective C calls for ANS staff to determine which ANS and pathways of introduction should be targeted for management actions. In this prioritization process, state staff should consider species' colonization potential and ecological impacts and how climate change may affect these processes. Increased precipitation, for example, may make habitat more suitable for certain ANS, increasing their colonization potential. Increased precipitation could also make the habitat less suitable for certain native species, allowing ANS to more easily become established.

C.4.4.3. Early Detection and Rapid Response, Control, and Management

Objective D discusses evaluating management alternatives. Strategies under this Objective should consider how climate change may alter the efficacy of some management strategies. For example, biocontrol agents may be affected by climate change, making them less effective or possibly allowing the biocontrol agents themselves to become invasive.

C.4.4.4. Research

As part of the Management Plan's effort to fill information gaps (Objective B), Strategy B1 calls for surveys of existing ANS within the Lake Champlain Basin, ANS with the potential to enter the basin, and pathways of introduction. Research on species that could enter the basin should consider how climate change effects may influence these species and provide opportunities for species not yet considered a threat under current climate conditions.

C.5. MIDWEST REGION AQUATIC NUISANCE SPECIES ACTION PLAN

C.5.1. GENERAL DESCRIPTION OF THE MIDWEST REGION'S PLAN

The Midwest Region Aquatic Nuisance Species (ANS) Action Plan is a product of the ANS Regulations and Enforcement Workshop, held in Indianapolis, Indiana on December 12, 2004 (available at http://www.protectyourwaters.net/ansreport/MWActionPlan.pdf). The goal of the Enforcement Workshop was to develop a regional assessment of ANS issues that can be addressed through increased coordination and communication among entities in the region. The Action Plan identifies a list of highest priority issues for immediate action, including funding, preventing new introductions and spread of ANS, early detection and rapid response, coordination, research, and public education. Each priority issue is accompanied by action items at the regional level and a procedure for implementation.

C.5.2. CLIMATE CHANGE AND INVASIVE SPECIES IN THE MIDWEST REGION

Climate models project temperatures increases of 5 to 10°F (2.8 to 5.5°C) and precipitation increases of 10 to 30% in the Midwest region by 2100 (Easterling and Karl, 2001). Increasing temperatures are expected to increase evaporation, triggering a soil moisture deficit, reduction in lake and river water levels, and diminished groundwater recharge. Models project lake level losses of up to 5 feet (1.5 meters) in the Great Lakes. In smaller lakes and rivers, increased evaporation will cause reduced flows, exacerbating water quality issues (Magnuson et al., 1997; Easterling and Karl, 2001). Heavy precipitation events are also expected to increase, washing nutrients and runoff from urban, agriculture, and construction sites into waterways (Magnuson et al., 1997).

Warmer water temperatures may make waterways more vulnerable to ANS invasions; aquatic ecosystems in the Midwest may experience a shift from coldwater fish species such as trout to warmer water species such as bass and catfish (Robillard and Fox, 2006; Jacobs et al., 2001).

C.5.3. THE MIDWEST REGION PLAN'S CURRENT INTEGRATION OF CLIMATE CHANGE

Table C-4 summarizes how the Midwest Region ANS Action Plan addresses and incorporates the projected effects of climate change. While the Action Plan does not explicitly address climate change, it does provide opportunity to increase attention to regional ANS issues, which include changing conditions.

Table C-4. Assessment of the Midwest Region Aquatic Nuisance Species Action Plan

Aspects of plan that may incorporate climate change	Score
Understanding and incorporating potential impacts resulting from climate change 0 = no; 1 = briefly mentions; 2 = includes general discussion; 3 = includes quantitative info and/or specific examples	
Plan specifically mentions climate change	0
Plan acknowledges climatic boundaries of species	0
Plan demonstrates understanding of species and/or ecosystem sensitivity to changing conditions	0
Plan identifies research on the potential effects of species responding to changing conditions	0
Plan acknowledges regional differences in expected climate changes	0
Capacity to adapt to changing conditions 0 = no; 1 = implicitly (i.e., includes goals and strategies that can be used to account for changing conditions, but does not specify changing conditions as part of their purpose); 2 = yes, explicitly, in passing; 3 = yes, explicitly, and specifies associated goals and/or action items	
Plan accounts for changing conditions in its leadership and coordination goals and strategies	1
Plan accounts for changing conditions in its prevention goals and strategies	0
Plan accounts for changing conditions in its early detection and rapid response goals and strategies	0
Plan accounts for changing conditions in its control and management goals and strategies	0
Plan accounts for changing conditions in its restoration goals and strategies	0
Plan accounts for changing conditions in its research goals and strategies	0
Plan accounts for changing conditions in its information management goals and strategies	0
Plan accounts for changing conditions in its education and public awareness goals and strategies	0
Monitoring strategies 0 = no; 1 = yes, briefly mentions; 2 = yes, but unclear how information will be used; 3 = yes, and specifies associated goals and/or action items	
Plan includes strategy to monitor for changing conditions	0
Plan includes strategy to utilize monitoring data	0
Plan includes strategy for managing/updating monitoring data	0
Revision 0 = no; 1 = yes, in passing; 2 = yes, and includes qualitative description; 3 = yes, and includes timeline and/or benchmarks for doing so	
Plan includes strategy for updating and incorporating new information	2
Funding 0 = no; 1 = a source is specified for a portion of the required funding; 2 = a source is specified for a portion of the required funding along with strategies for obtaining remaining funding; 3 = a source is specified for 100% of required funding	
Plan identifies dedicated funding source for implementation (includes suggested funding sources)	0
Total score:	**3**

C.5.4. INCORPORATING CLIMATE CHANGE INFORMATION

Climate change will affect the Midwest region and its water resources throughout the next decades. The ANS Action Plan should incorporate climate change effects so that management activities can better respond to changing conditions and remain effective over time.

Several of the Strategic Actions listed in the Action Plan could incorporate climate change considerations to adapt management activities in the Midwest region.

C.5.4.1. Leadership and Coordination

Issue 6 of the Action Plan addresses partnerships and coordination, including involving external organizations such as nongovernmental organizations and industry in promoting ANS policy and agendas. These provisions call for expanding coordination and attention to regional ANS issues. Coordination on the regional level offers the opportunity for states to share information on climate-related ANS issues, including expanding ANS ranges, changing conditions, and adapting management to address changing conditions.

C.5.4.2. Prevention

Issue 2 relates to preventing new introductions and the spread of ANS, including the development of a method for evaluating species to determine whether they should be allowed to enter a state or the country. Issue 9 involves coordinating and communicating regulated species lists among the states and developing federal guidance for uniformity between states. Efforts to assess risk and develop species lists could also account for the effect of climate change on ecosystem vulnerability to ANS invasion and expanding species ranges and vectors.

C.5.4.3. Early Detection and Rapid Response, Control, and Management

Issue 12 addresses developing tools for control and management of established ANS. These tools should incorporate climate change effects to ensure long-term effectiveness.

C.5.4.4. Research

Issue 4 promotes the compilation and use of economic data to generate interest in, and support for, ANS issues from sport fishing and hunting communities. This effort includes researching ANS impacts, producing economic data on ANS impacts and educating policymakers and the public. Analyses of ecologic and economic impacts from ANS should also take climate change effects into account and be communicated to the public.

C.6. NORTHEAST REGION AQUATIC NUISANCE SPECIES ACTION PLAN

C.6.1. GENERAL DESCRIPTION OF THE NORTHEAST REGION PLAN

The Northeast Region Aquatic Nuisance Species (ANS) Action Plan is based on the results of the ANS Regulations and Enforcement Workshops held in Ocean City, Maryland and Atlantic City, New Jersey in 2004 (available at http://www.protectyourwaters.net/ansreport/NEActionPlan.pdf). The Enforcement Workshops were coordinated by the Association of Fish and Wildlife Agencies. The Action Plan, which focuses on regional cooperation and coordination, identifies the following high priority issues: funding for state and regional ANS management actions, coordination and communication of ANS lists, development of mechanisms for tracking and controlling Internet sales and other shipments, and development of ANS screening and risk assessment tools. Each priority issue is accompanied by corresponding priority issues at the regional level and a process for implementation.

C.6.2. CLIMATE CHANGE AND INVASIVE SPECIES IN THE NORTHEAST REGION

Climate models project that minimum winter temperatures in the Northeast will increase from 5 to 9.5°F (2.9 to 5.3°C) by 2099 depending on the model (Hayhoe et al., 2007). Precipitation projections vary, ranging from increases of up to 30% in winter months by 2099 to decreases or small changes in summer months (Hayhoe et al., 2007). Temperature and precipitation changes are likely to increase water temperature in the region's waterbodies, which will affect species survival, dissolved oxygen levels, and nutrient content (Hayhoe et al., 2007).

Increased water temperatures may increase the region's susceptibility to invasion by ANS. In Massachusetts, for example, water hyacinth may be limited in its establishment by its inability to overwinter in the New England climate. However, as water temperatures increase, water hyacinth communities that become established during the summer may be able to survive the winter (Massachusetts Department of Natural Resources, 2003).

C.6.3. THE NORTHEAST REGION PLAN'S CURRENT INTEGRATION OF CLIMATE CHANGE

Table C-5 summarizes how the Northeast Region ANS Action Plan addresses and incorporates the projected effects of climate change. The Northeast Region's Action Plan does not address climate change effects or changing environmental conditions in general.

Table C-5. Assessment of the Northeast Region Aquatic Nuisance Species Action Plan

Aspects of plan that may incorporate climate change	Score
Understanding and incorporating potential impacts resulting from climate change 0 = no; 1 = briefly mentions; 2 = includes general discussion; 3 = includes quantitative info and/or specific examples	
Plan specifically mentions climate change	0
Plan acknowledges climatic boundaries of species	0
Plan demonstrates understanding of species and/or ecosystem sensitivity to changing conditions	0
Plan identifies research on the potential effects of species responding to changing conditions	0
Plan acknowledges regional differences in expected climate changes	0
Capacity to adapt to changing conditions 0 = no; 1 = implicitly (i.e., includes goals and strategies that can be used to account for changing conditions, but does not specify changing conditions as part of their purpose); 2 = yes, explicitly, in passing; 3 = yes, explicitly, and specifies associated goals and/or action items	
Plan accounts for changing conditions in its leadership and coordination goals and strategies	0
Plan accounts for changing conditions in its prevention goals and strategies	0
Plan accounts for changing conditions in its early detection and rapid response goals and strategies	0
Plan accounts for changing conditions in its control and management goals and strategies	0
Plan accounts for changing conditions in its restoration goals and strategies	0
Plan accounts for changing conditions in its research goals and strategies	0
Plan accounts for changing conditions in its information management goals and strategies	0
Plan accounts for changing conditions in its education and public awareness goals and strategies	0
Monitoring strategies 0 = no; 1 = yes, briefly mentions; 2 = yes, but unclear how information will be used; 3 = yes, and specifies associated goals and/or action items	
Plan includes strategy to monitor for changing conditions	0
Plan includes strategy to utilize monitoring data	0
Plan includes strategy for managing/updating monitoring data	0
Revision 0 = no; 1 = yes, in passing; 2 = yes, and includes qualitative description; 3 = yes, and includes timeline and/or benchmarks for doing so	
Plan includes strategy for updating and incorporating new information	0
Funding 0 = no; 1 = a source is specified for a portion of the required funding; 2 = a source is specified for a portion of the required funding along with strategies for obtaining remaining funding; 3 = a source is specified for 100% of required funding	
Plan identifies dedicated funding source for implementation	0
Total score:	**0**

C.6.4. INCORPORATING CLIMATE CHANGE INFORMATION

Climate change will affect the Northeast region and its water resources throughout the next decades. The ANS Action Plan should incorporate climate change effects so that management activities can better respond to changing conditions and remain effective over time.

Several of the Strategic Actions listed in the Action Plan could incorporate climate change considerations to adapt management activities in the Northeast region.

C.6.4.1.   Leadership and Coordination

Issue 2 involves improving regulation and enforcement through the coordination of regulated species lists among states in the region. The Action Plan encourages states to develop a shared methodology and scientific criteria for creating lists and to implement interstate agreements on the development of regional restricted species lists. The development of list criteria should include climate change information that affects ANS' invasive potential and ecosystem alterations such as increased stream flows.

C.6.4.2.   Prevention

Issue 4 addresses the improvement of ANS screening and risk assessment tools. Risk assessment of ANS should also account for the Northeast's increased sensitivity to ANS invasion as a result of climate change. Consideration of climate change effects could improve risk determinations.

C.6.4.3.   Early Detection and Rapid Response, Control, and Management

One of the actions under Issue 1 encourages the development of early detection and rapid response plans for the states. These plans should also consider the effects from climate change on ANS pathways, establishment and spread in order to develop plans that remain effective under changing environmental conditions.

C.6.4.4.   Research

Appendix D encourages states in the region to develop support research that advances control, eradication, monitoring, and prevention of ANS. These research initiatives should also consider the effects of climate change on each of these activities to ensure that implementation plans are written to take changing conditions into account.

C.7. SOUTHEAST REGION AQUATIC NUISANCE SPECIES ACTION PLAN

C.7.1. GENERAL DESCRIPTION OF THE SOUTHEAST REGION'S PLAN

The Southeast Region Aquatic Nuisance Species (ANS) Action Plan is a product of the ANS Regulations and Enforcement Workshop, held in Hilton Head, South Carolina in 2004 (available at http://www.protectyourwaters.net/ansreport/SEActionPlan.pdf). The resulting Action Plan was coordinated by the International Association of Fish and Wildlife Agencies. The Action Plan identifies six issues for immediate action: (1) locate funding for ANS management; (2) coordinate ANS lists among states; (3) describe regulatory authorities; (4) coordinate regional ANS management; (5) generate support from external organizations by developing economic impact information; and (6) enhance ANS detection and rapid response capabilities. The issues are accompanied by priority actions and a step-by-step process for implementation.

C.7.2. CLIMATE CHANGE AND INVASIVE SPECIES IN THE SOUTHEAST REGION

Most climate change models project warmer temperatures in the Southeast and some models project a 20% increase in precipitation (Burkett et al., 2001; Mulholland et al., 1997). Pollution is already prevalent in waterbodies in the Southeast, and increased precipitation and more frequent extreme precipitation events may flush more contaminated run-off into waterbodies. Higher temperatures will also reduce dissolved oxygen levels (Mulholland et al., 1997). Sea level rise is also projected for the region, leading to flooding and erosion, increasing storm surges, coastal wetland loss that reduces habitat for cool water fisheries, and the conversion of freshwater inflows into estuaries (Burkett et al., 2001; Mulholland et al., 1997).

These ecosystem changes may benefit ANS that are more tolerant of poor water quality than native species. In coastal regions, ANS may become established as freshwater areas become increasingly saline. New conditions may also make native species more susceptible to viral ANS. Largemouth Bass Virus (LMBV) has been found in fish in Southeastern states. The virus has a higher mortality rate in bass that are living in stressful conditions, including warm-water temperatures and poor water quality (Inendino et al., 2005; Grant et al., 2003). As climate change increases these two conditions, the region may experience LMBV-related fish kills.

C.7.3. THE SOUTHEAST REGION PLAN'S CURRENT INTEGRATION OF CLIMATE CHANGE

Table C-6 summarizes how the Southeast Region ANS Action Plan addresses and incorporates the projected effects of climate change. While the Southeast Region's Action Plan does not address climate change, it does provide opportunity to incorporate changing conditions as they arise into its efforts to coordinate ANS management at the regional level.

Table C-6. Assessment of the Southeast Region Aquatic Nuisance Species Action Plan

Aspects of plan that may incorporate climate change	Score
Understanding and incorporating potential impacts resulting from climate change 0 = no; 1 = briefly mentions; 2 = includes general discussion; 3 = includes quantitative info and/or specific examples	
Plan specifically mentions climate change	0
Plan acknowledges climatic boundaries of species	0
Plan demonstrates understanding of species and/or ecosystem sensitivity to changing conditions	0
Plan identifies research on the potential effects of species responding to changing conditions	0
Plan acknowledges regional differences in expected climate changes	0
Capacity to adapt to changing conditions 0 = no; 1 = implicitly (i.e., includes goals and strategies that can be used to account for changing conditions, but does not specify changing conditions as part of their purpose); 2 = yes, explicitly, in passing; 3 = yes, explicitly, and specifies associated goals and/or action items	
Plan accounts for changing conditions in its leadership and coordination goals and strategies	1
Plan accounts for changing conditions in its prevention goals and strategies	0
Plan accounts for changing conditions in its early detection and rapid response goals and strategies	0
Plan accounts for changing conditions in its control and management goals and strategies	0
Plan accounts for changing conditions in its restoration goals and strategies	0
Plan accounts for changing conditions in its research goals and strategies	0
Plan accounts for changing conditions in its information management goals and strategies	0
Plan accounts for changing conditions in its education and public awareness goals and strategies	0
Monitoring strategies 0 = no; 1 = yes, briefly mentions; 2 = yes, but unclear how information will be used; 3 = yes, and specifies associated goals and/or action items	
Plan includes strategy to monitor for changing conditions	0
Plan includes strategy to utilize monitoring data	0
Plan includes strategy for managing/updating monitoring data	0
Revision 0 = no; 1 = yes, in passing; 2 = yes, and includes qualitative description; 3 = yes, and includes timeline and/or benchmarks for doing so	
Plan includes strategy for updating and incorporating new information	0
Funding 0 = no; 1 = a source is specified for a portion of the required funding; 2 = a source is specified for a portion of the required funding along with strategies for obtaining remaining funding; 3 = a source is specified for 100% of required funding	
Plan identifies dedicated funding source for implementation	0
Total score:	**1**

C.7.4. INCORPORATING CLIMATE CHANGE INFORMATION

In light of the range of climate change effects expected in the Southeast region, especially the effects of rising sea levels on coastal marshes, the Action Plan should address climate change in its ANS management activities. Several actions in the Action Plan could incorporate climate change considerations to adapt management activities in the Southeast Region.

C.7.4.1. Leadership and Coordination

Issue 4 suggests regional ANS management coordination. This requires the identification and discussion of differences among the states about which species are of highest concern. Risks are to be ranked to determine regional and state management strategies. Risks assessments that consider the effects of changing conditions on ANS' invasive potential may account for ANS threats more effectively.

C.7.4.2. Prevention

Although the Action Plan does not explicitly address prevention, Issue 2 in the Action Plan calls for the coordination of ANS lists. The development of these lists should consider climate change effects on species introductions and distributions, and coordination of lists will ensure that species that may easily become established under changing conditions can be addressed by neighboring states.

C.7.4.3. Early Detection and Rapid Response, Control, and Management

Issue 6 encourages the enhancement of detection and rapid response capabilities. Species detection systems should also consider the effects of climate change, in order to determine likely vectors and vulnerable ecosystems.

C.7.4.4. Research

The Action Plan suggests that member agencies increase consistency among states by developing screening and risk assessment tools. These efforts should be designed to account for projected effects of climate change. Changing conditions are likely to influence the ability of ANS to become established, as well as ecosystem vulnerability to invasions.

C.8. WESTERN REGION AQUATIC NUISANCE SPECIES ACTION PLAN

C.8.1. GENERAL DESCRIPTION OF THE WESTERN REGION'S PLAN

The Western Region Aquatic Nuisance Species Action Plan is based on discussions and findings from the ANS Regulations and Enforcement Workshop, held in 2004 in Sun Valley, Idaho (available at http://www.protectyourwaters.net/ansreport/WestActionPlan.pdf). The Enforcement Workshop was coordinated by the International Association of Fish and Wildlife Agencies. The resulting Action Plan identifies nine priority issues that address (1) funding ANS management programs; (2) training enforcement officers on species identification; (3) identifying entities to promote ANS policies; (4) developing screening and risk assessment tools; (5) developing mechanisms to track and control internet sales; (6) developing rapid response capabilities; (7) identifying a regional structure to address ANS; (8) developing ANS lists; and (9) communicating about ANS laws. Each of these priority issues is accompanied by a list of actions items.

C.8.2. CLIMATE CHANGE AND INVASIVE SPECIES IN THE WESTERN REGION

In the western states, climate change models project air temperature increases of 3 to 4°F (1.6 to 2.2°C) by the 2030s and 8 to 11°F (4.4 to 6.1°C) by the 2090s (Parson et al., 2001). In California annual temperatures may increase by 4 to 10°F (2.3 to 5.8°C) between 2070 and 2090 (Hayhoe et al., 2006). Climate change models also project an increase in winter precipitation and a decrease in summer precipitation, with an annual average precipitation increase from 10 to 30% (Parson et al., 2001), although California modeling predicts an annual decrease in precipitation (Hayhoe et al., 2006). These precipitation changes will lead to earlier melting of snow pack, increased flooding in rain-fed rivers, and summer water shortages. Climate change models also project more extreme wet and dry years (Parson et al., 2001). Along the coast rising sea levels threaten coastal wetlands and the species they support.

Climate change may also facilitate the spread and establishment of aquatic invasive species in the west. For example, water hyacinth in Washington is thought to be limited in its ability to become established because of the state's cold winters (Washington Department of Fish and Wildlife, 2001). As increasing temperatures warm water bodies in the region, the waters of Washington may be more suitable to water hyacinth, allowing the plant to become widely established.

C.8.3. THE WESTERN REGION PLAN'S CURRENT INTEGRATION OF CLIMATE CHANGE

Table C-7 summarizes how the Western Region ANS Action Plan addresses and incorporates the projected effects of climate change. The Action Plan does not address climate change, but provides the opportunity to incorporate changing conditions into action items.

Table C-7. Assessment of the Western Region Aquatic Nuisance Species Action Plan

Aspects of plan that may incorporate climate change	Score
Understanding and incorporating potential impacts resulting from climate change 0 = no; 1 = briefly mentions; 2 = includes general discussion; 3 = includes quantitative info and/or specific examples	
Plan specifically mentions climate change	0
Plan acknowledges climatic boundaries of species	0
Plan demonstrates understanding of species and/or ecosystem sensitivity to changing conditions	0
Plan identifies research on the potential effects of species responding to changing conditions	0
Plan acknowledges regional differences in expected climate changes	0
Capacity to adapt to changing conditions 0 = no; 1 = implicitly (i.e., includes goals and strategies that can be used to account for changing conditions, but does not specify changing conditions as part of their purpose); 2 = yes, explicitly, in passing; 3 = yes, explicitly, and specifies associated goals and/or action items	
Plan accounts for changing conditions in its leadership and coordination goals and strategies	0
Plan accounts for changing conditions in its prevention goals and strategies	0
Plan accounts for changing conditions in its early detection and rapid response goals and strategies	1
Plan accounts for changing conditions in its control and management goals and strategies	0
Plan accounts for changing conditions in its restoration goals and strategies	0
Plan accounts for changing conditions in its research goals and strategies	0
Plan accounts for changing conditions in its information management goals and strategies	0
Plan accounts for changing conditions in its education and public awareness goals and strategies	0
Monitoring strategies 0 = no; 1 = yes, briefly mentions; 2 = yes, but unclear how information will be used; 3 = yes, and specifies associated goals and/or action items	
Plan includes strategy to monitor for changing conditions	0
Plan includes strategy to utilize monitoring data	0
Plan includes strategy for managing/updating monitoring data	0
Revision 0 = no; 1 = yes, in passing; 2 = yes, and includes qualitative description; 3 = yes, and includes timeline and/or benchmarks for doing so	
Plan includes strategy for updating and incorporating new information	0
Funding 0 = no; 1 = a source is specified for a portion of the required funding; 2 = a source is specified for a portion of the required funding along with strategies for obtaining remaining funding; 3 = a source is specified for 100% of required funding	
Plan identifies dedicated funding source for implementation	0
Total score:	**1**

C.8.4. INCORPORATING CLIMATE CHANGE INFORMATION

Climate change will affect the Western region and its water resources throughout the next decades. The ANS Action Plan should incorporate climate change effects so that management activities can better respond to changing conditions and remain effective over time. Several of the Strategic Actions listed in the Action Plan could incorporate climate change considerations to adapt management activities in the Western region.

C.8.4.1. Leadership and Coordination

Issue 7 of the Action Plan proposes enhancing the regional organizational structure for addressing ANS, and Issue 8 proposes coordination of ANS lists among states and the clarification of definitions for regulation and policy. States in the region could incorporate climate change information into this structure by planning communication strategies and sharing information on the regional effects of climate change, as well as changes in habitats and species ranges.

C.8.4.2. Prevention

The Action Plan encourages the development of screening and risk assessment tools for ANS identification and management (Issue 4). This includes the identification of groups of species and ecosystems where the greatest threat exists and the prioritization of these risks. Risk assessment of ANS could also account for the projected effects of climate change. Changing conditions may alter some habitats in favor of some ANS, thereby facilitating establishment.

C.8.4.3. Early Detection and Rapid Response, Control, and Management

Issue 6 of the Action Plan outlines the development of a regional rapid response strategy. This section calls for more communication and cooperation among states and awareness of what is happening in other states, the region, and the nation. Attention to how ecosystems may be affected by climate change and how ANS are responding to changing conditions in other states and regions can help western states be better prepared to address these effects.

C.8.4.4. Research

The Action Plan identifies the development of risk assessment and screening tools as a need in Issue 4 and lists research as one of the specific actions. Risk assessments and screening tools should incorporate climate change effects in order to provide more accurate analyses of ANS threats.

REFERENCES

Burkett, V; Kusler, J. (2000) Climate change: potential impacts and interactions in wetlands of the United States. J Am Water Res Assoc 36(2):313–320.

Burkett, V; Ritschard, R; McNulty, S; et al. (2001) Potential consequences of climate variability and change for the southeastern United States. In: Climate change impacts on the United States: the potential consequences of climate variability and change. Report for the Global Change Research Program. Cambridge, England: Cambridge University Press; pp. 137–164. Available online at http://www.usgcrp.gov/usgcrp/Library/nationalassessment/05SE.pdf.

Easterling, DR; Karl, TR. (2001) Potential consequences of climate variability and change for the midwestern United States. In: Climate change impacts on the United States: the potential consequences of climate variability and change. Report for the US Global Change Research Program. National Assessment Synthesis Team. Cambridge, United Kingdom: Cambridge University Press; pp. 167–188. Available online at http://www.gcrio.org/NationalAssessment/6MW.pdf [accessed May 1, 2007].

Gibson, CA; Meyer, JL; Poff, NL; et al. (2005) Flow regime alterations under changing climate in two river basins: implications for freshwater ecosystems. River Res Appl 21:849–864.

Grant, EC; Philipp, DP; Inendino, KR; et al. (2003) Effects of temperature on the susceptibility of largemouth bass to largemouth bass virus. J Aquat Anim Health 15(3):215–220.

Hayhoe, K; Cayan, D; Field, CB; et al. (2006) Emissions pathways, climate change, and impacts on California. Proc Natl Acad Sci 101(34):12422–12427.

Hayhoe, K; Wake, CP; Huntington, TG; et al. (2007) Past and future changes in climate and hydrological indicators in the US Northeast. Clim Dyn 28:381–407.

Inendino, KR; Grant, EC; Philipp, DP; et al. (2005) Effects of factors related to water quality and population density on the sensitivity of juvenile largemouth bass to mortality induced by viral infection. J Aquat Anim Health 17(4):304–314.

Jacobs, B; Adams, B; Gleick, P. (2001) Potential consequences of climate variability and change for the water resources of the United States. In: Climate change impacts on the United States: the potential consequences of climate variability and change. Report for the Global Change Research Program. Cambridge, England: Cambridge University Press; pp. 405–435. Available online at http://www.usgcrp.gov/usgcrp/Library/nationalassessment/14Water.pdf.

Magnuson, JJ; Webster, KE; Assel, RA; et al. (1997) Potential effects of climate changes on aquatic systems: Laurentian Great Lakes and Precambrian Shield Region. Hydrol Processes 11(8):825–871.

Massachusetts Department of Natural Resources. (2003) Potential invader. Water hyacinth, an exotic aquatic plant. (*Eichhornia crassipes*). Available online at http://www.mass.gov/dcr/waterSupply/lakepond/factsheet/water%20hyacinth.pdf [accessed Apr. 26, 2007].

Mearns, LO; Giorgi, F., McDaniel, L., et al. (2003) Climate scenarios for the southeastern U.S. based on GCM and regional model simulations. Clim Change 60(1–2):7–35.

Mulholland, PJ; Best, GR; Coutant, CC; et al. (1997) Effects of climate change on freshwater ecosystems of the south-eastern United States and the Gulf Coast of Mexico. Hydrol Processes 11(8):949–970.

Parson, EA; Mote, PW; Hamlet, A; et al. (2001) Potential consequences of climate variability and change for the Pacific Northwest. In: Climate change impacts on the United States: the potential consequences of climate variability and change. Report for the US Global Change Research Program. National Assessment Synthesis Team. Cambridge, United Kingdom: Cambridge University Press; pp. 247–280. Available online at http://www.usgcrp.gov/usgcrp/Library/nationalassessment/09PNW.pdf [accessed May 1, 2007].

Robillard, MM; Fox, MG. (2006) Historical changes in abundance and community structure of warmwater piscivore communities associated with changes in water clarity, nutrients, and temperature. Can J Fish Aquat Sci 63(4):798–809.

National Assessment Synthesis Team (NAST). (2001) Climate change impacts on the United States: the potential consequences of climate variability and change. Cambridge University Press, Cambridge UK.

U.S. EPA (U.S. Environmental Protection Agency). (2000) Aquatic nuisance species. In: Lake Superior Lakewide Management Plan; Chapter 10. Available online at http://gleams.altarum.org/glwatershed/lamps/lakesuperior/2000/LS%20chapter%2010.pdf [accessed Apr. 26, 2007].

Washington Department of Fish and Wildlife. (2001) Aquatic nuisance species management plan. The Washington Aquatic Nuisance Species Coordinating Committee. Available online at http://wdfw.wa.gov/fish/ans/2001ansplan.pdf [accessed Apr. 26, 2007].

APPENDIX D

COMPLETE CRITERIA AND SCORING FOR STATE PLAN CONSIDERATION OF CLIMATE CHANGE AND/OR CHANGING CONDITIONS

We reviewed state aquatic invasive species (AIS) management plans, where available, and assessed how they address climate change specifically, as well as how they provide generally for adaptation of management strategies and actions under changing conditions. In total, we reviewed 25 state plans, including 23 AIS-specific plans and 2 general invasive species management plans with a significant AIS focus. Categories of assessment include the following:

(1) How the plan addresses potential impacts resulting from climate change

(2) How the plan demonstrates capacity to adapt to changing conditions

(3) How the plan provides monitoring strategies

(4) Whether and to what extent the plan provides for periodic revision and update

(5) Whether and to what extent the plan describes funding sources/strategies for plan implementation

The sections below correspond to each assessment category, followed by a section presenting each state plan's summary score and rank among assessed states.

D.1. HOW THE PLAN ADDRESSES POTENTIAL IMPACTS RESULTING FROM CLIMATE CHANGE

Table D-1. Understanding and incorporating potential impacts resulting from climate change

SCORING:
0 = no; 1 = briefly mentions; 2 = includes general discussion; 3 = includes quantitative info and/or specific examples

	Plan specifically mentions climate change	Plan acknowledges climatic boundaries of species	Plan demonstrates understanding of species and/or ecosystem sensitivity to changing conditions	Plan identifies research on the potential effects of species responding to changing conditions	Plan acknowledges regional differences in expected climate changes
Alaska	0	2	2	0	0
Arizona	0	1	0	0	0
Connecticut	1	1	1	0	0
Hawaii	0	2	2	0	0
Idaho	0	1	0	0	0
Illinois	0	1	1	0	0
Indiana	0	3	0	0	0
Iowa	0	1	0	0	0
Kansas	0	0	0	0	0
Louisiana	0	3	3	0	0
Maine	1	2	2	0	0
Massachusetts	1	3	1	0	0
Michigan	0	1	0	0	0
Missouri	0	3	0	0	0
Montana	0	0	1	0	0
New York	0	0	0	0	0
North Dakota	0	2	1	0	0
Ohio	0	1	0	0	0
Oregon	0	3	0	0	0
Pennsylvania	0	0	0	0	0
South Carolina	0	0	0	0	0
Texas	0	0	0	0	0
Virginia	2	1	1	0	0
Washington	0	3	0	0	0
Wisconsin	0	1	0	0	0

D.2. HOW THE PLAN DEMONSTRATES CAPACITY TO ADAPT TO CHANGING CONDITIONS

Table D-2. Capacity to adapt to changing conditions

SCORING:
0 = no; 1 = implicitly (i.e., includes goals and strategies that can be used to account for changing conditions but does not specify changing conditions as part of their purpose); 2 = yes, explicitly, in passing; 3 = yes, explicitly, and specifies associated goals and/or action items

	Plan accounts for changing conditions in its goals and strategies for…							
	…leadership and coordination	…prevention	…EDRR	…control and management	…restoration	…research	…information management	…education and public awareness
Alaska	1	1	0	0	0	1	0	1
Arizona	0	1	0	0	0	1	0	0
Connecticut	1	1	0	1	0	1	0	0
Hawaii	0	0	1	0	0	1	0	1
Idaho	0	0	0	0	0	0	0	0
Illinois	0	0	0	0	0	2	0	0
Indiana	0	1	1	0	0	0	0	0
Iowa	0	0	0	0	0	0	0	0
Kansas	0	0	0	0	0	3	0	0
Louisiana	0	0	1	0	0	0	0	0
Maine	0	0	0	0	0	0	0	0
Massachusetts	0	0	0	0	0	0	0	0
Michigan	0	0	0	0	0	0	0	0
Missouri	0	0	0	0	0	0	0	0
Montana	0	0	0	0	0	3	0	0
New York	0	0	0	0	0	0	0	0
North Dakota	0	0	0	0	0	1	0	0
Ohio	0	0	0	0	0	2	0	0
Oregon	0	0	0	0	0	0	0	0
Pennsylvania	0	0	0	0	0	0	0	0
South Carolina	0	0	0	0	0	0	0	0
Texas	0	0	0	0	0	0	0	0
Virginia	0	0	0	0	0	0	0	0
Washington	0	0	0	0	0	2	0	1
Wisconsin	0	0	0	0	0	0	0	0

D.3. HOW THE PLAN PROVIDES MONITORING STRATEGIES

Table D-3. Monitoring strategies

SCORING: 0 = no; 1 = yes, briefly mentions; 2 = yes, but unclear how information will be used; 3 = yes, and specifies associated goals and/or action items		
	Plan includes strategy to monitor for changing conditions	**Plan includes strategy to utilize monitoring data**	**Plan includes strategy for managing/updating monitoring data**
Alaska	0	2	3
Arizona	0	3	0
Connecticut	0	1	1
Hawaii	0	3	3
Idaho	0	0	0
Illinois	0	1	1
Indiana	0	3	0
Iowa	0	3	0
Kansas	0	3	3
Louisiana	0	3	0
Maine	1	0	0
Massachusetts	0	3	0
Michigan	0	3	0
Missouri	0	3	3
Montana	0	0	0
New York	0	3	0
North Dakota	0	1	1
Ohio	0	1	0
Oregon	0	0	0
Pennsylvania	0	0	0
South Carolina	0	0	1
Texas	0	2	3
Virginia	0	0	0
Washington	0	3	3
Wisconsin	0	3	0

D.4. WHETHER, AND TO WHAT EXTENT, THE PLAN PROVIDES FOR PERIODIC REVISION AND UPDATE

Table D-4. Revision

SCORING:
0 = no; 1 = yes, in passing; 2 = yes, and includes qualitative description;
3 = yes, and includes timeline and/or benchmarks for doing so

	Plan includes strategy for updating and incorporating new information
Alaska	2
Arizona	0
Connecticut	1
Hawaii	1
Idaho	0
Illinois	0
Indiana	3
Iowa	2
Kansas	3
Louisiana	0
Maine	3
Massachusetts	0
Michigan	0
Missouri	0
Montana	3
New York	0
North Dakota	1
Ohio	1
Oregon	3
Pennsylvania	2
South Carolina	2
Texas	0
Virginia	3
Washington	3
Wisconsin	1

D.5. WHETHER, AND TO WHAT EXTENT, THE PLAN DESCRIBES FUNDING SOURCES/STRATEGIES FOR PLAN IMPLEMENTATION

Table D-5. Funding

SCORING: 0 = no; 1 = a source is specified for a portion of the required funding; 2 = a source is specified for a portion of the required funding along with strategies for obtaining remaining funding; 3 = a source is specified for 100% of required funding	
	Plan identifies dedicated funding source for implementation
Alaska	1
Arizona	0
Connecticut	2
Hawaii	0
Idaho	0
Illinois	0
Indiana	1
Iowa	2
Kansas	2
Louisiana	2
Maine	0
Massachusetts	2
Michigan	0
Missouri	3
Montana	2
New York	1
North Dakota	2
Ohio	0
Oregon	3
Pennsylvania	1
South Carolina	3
Texas	0
Virginia	0
Washington	2
Wisconsin	3

D.6. TOTAL SCORING

The table below sums each state plan's total score for the 5 assessment categories and presents the plan's rank among the 25 states.

Table D-6. Total score and ranking for 25 state plans' consideration of climate change and/or provisions for adaptation of strategies and actions under changing conditions

	D1. Understanding and incorporating potential impacts resulting from climate change	D2. Capacity to adapt to changing conditions	D3. Monitoring strategies	D4. Plan includes strategy for updating and incorporating new information	D5. Plan identifies dedicated funding source for implementation	Total Score (D1-D5)	Rank among 25 states
Alaska	4	4	5	2	1	16	2
Arizona	1	2	3	0	0	6	9
Connecticut	3	4	2	1	2	12	4
Hawaii	4	3	6	1	0	14	3
Idaho	1	0	0	0	0	1	13
Illinois	2	2	2	0	0	6	9
Indiana	3	2	3	3	1	12	4
Iowa	1	0	3	2	2	8	7
Kansas	0	3	6	3	2	14	3
Louisiana	6	1	3	0	2	12	4
Maine	5	0	1	3	0	9	6
Massachusetts	5	0	3	0	2	10	5
Michigan	1	0	3	0	0	4	11
Missouri	3	0	6	0	3	12	4
Montana	1	3	0	3	2	9	6
New York	0	0	3	0	1	4	11
North Dakota	3	1	2	1	2	9	6
Ohio	1	2	1	1	0	5	10
Oregon	3	0	0	3	3	9	6
Pennsylvania	0	0	0	2	1	3	12
South Carolina	0	0	1	2	3	6	9
Texas	0	0	5	0	0	5	10
Virginia	4	0	0	3	0	7	8
Washington	3	3	6	3	2	17	1
Wisconsin	1	0	3	1	3	8	7

APPENDIX E

MODELS FOR INVASIVE SPECIES INTRODUCTION, ESTABLISHMENT, SPREAD, AND INVASION

Models of invasive species introductions, distribution and spread, and establishment are key tools for both understanding the invasive species problem and designing effective prevention and control techniques. Numerous types of models have been developed. In many cases, authors recommend that conservation managers be cognizant of specific factors (e.g., species interactions, climatic factors, spread vectors) in ecosystem management. Some authors offer clear, ready-to-use models and strategies for conservation managers. Table E-1 below lists examples of models used to predict species invasions.

Although most invasive species' spread, distribution, and establishment models are not designed specifically to incorporate climate change variables, several approaches have been developed that do explicitly address climate change impacts on species distributions. These include bioclimatic envelope models, discriminant analyses, and logistic regression analyses (see table below—identified by an asterisk). Other modeling methods, such as ecological niche modeling, could be modified to integrate climate change variables.

Table E-1. Model and description

Comparative analysis. Ricciardi and Rasmussen (1998) use descriptive information to identify species with high invasion and impact potential. The three steps include (1) identifying donor regions and dispersal; (2) selecting potential invaders based on biological traits; and (3) using invasion history. The analysis identifies *Corophium* spp., Mysids, and *Clupeonella caspia* as possible Great Lakes-St. Lawrence River system invaders. The authors recommend focusing on monitoring and applying described guidelines, and developing an accessible electronic database of possible invaders.
Comparative analysis. Vermeij (1996) recommends an agenda for invasion biology. The author supports using a model or approach that is more comparative and systematic that involves a framework of scientific questions that need to be answered. The paper does not include a scientific study; thus, no results are provided. However, the author recommends comparing factors involved in the invasion process (arrival, establishment, and integration), participants, and outcomes at spatial and temporal scales. The author prefers multiple methodological approaches to address invasion biology.
Simple diffusion model. Buchan and Padilla (1999) use a simple diffusion model to predict zebra mussel spread by (1) comparing current pattern of zebra mussel invasion with estimates of boater movements, and (2) diffusion model data. The model can be used to estimate infrequent, long-distance boater movements to predict AIS invasion probability. Simple diffusion models are mathematical models that use variables such as population growth and density and velocity of the invasion front to predict temporal and spatial patterns. The authors recommend managers use the results to predict spread rates and patterns to use in developing management strategies for the Great Lakes. Efforts to curb or stop spread should focus on high frequency long-distance paths such as areas with high boating activity.
Diffusion. Vanderploeg et al. (2002) describe and predict dispersal patterns and ecological impacts of five invaders in the Great Lakes. The authors synthesize laboratory data on the ecology of Ponto-Caspian invasive species, on the patterns of dispersal, and on the impacts of the invasive species. This information is applied to areas of the Great Lakes and used to generate case studies on change due to invasion, the causes of invasion, and future predictions. The results show a mix of continuous and discontinuous dispersal. Hypothesized general attributes of invasive species are valuable to predict successful invaders but not for determining impacts. The authors recommend that additional research focus on benthic food webs to understand the primary impact of invaders.

Table E-1. Model and description (continued)

Diffusion. Suarez et al. (2001) use a stratified diffusion model to reconstruct invasion dynamics of the Argentine ant (*Linepithema humile*) because it uses more than one dispersal process. The authors use museum records, personal surveys, literature, and unpublished communications at three different scales (worldwide, regional, and local) to determine patterns of dispersal. The authors find that human-induced "jump-dispersal" plays an important role in invasion patterns after establishment. The authors recommend that control measures focus on new foci or preventing new foci. They state that identifying the range of long-distance jump-dispersal will help future modeling efforts, and reconstructing spatial scales of invasion dynamics may make strategies for management and eradication more effective.
Reaction-diffusion model. Grosholz (1996) uses a reaction-diffusion model to highlight differences in invasions between marine and terrestrial species describes population behavior at the population level. Use of the model requires the assumption of "random movement, continuous positive population growth, a homogenous environment, and no taxis or interspecific interactions." It provides insights on invasions at a broader scale (not individual scale). The results show that using data on one invasion may not be a good predictor of other conspecific invasions. The author concludes that invasions may not accurately predict invasions for conspecifics and that diffusion models are useful for predicting general invasion patterns but not for predicting spread rates for specific invasions.
Reaction-diffusion model. Lonsdale (1993) tests whether Skellam's model for areal spread describes *Mimosa pigra* invasion and finds that it does not. Skellam's model is continuous, deterministic, and requires users to assume (1) population increasing exponentially; (2) diffusing outward randomly; and (3) normally distributed distribution. The author finds that climatic conditions such as rainfall and flooding increase rate of spread. The author concludes that Skellam's model does not sufficiently describe the growth of *M. pigra* and that population dynamics of invasive species are relatively simple.
Discrete event simulation. Hill et al. (1998) use a discrete event simulation model to determine the development and colonization of the green alga *Caulerpa taxifolia*. The model accounts for sensitive physical and biological factors and assesses and predicts *C. taxifolia* propagation a various scales. In addition to simulating and predicting expansion, the model produces GIS maps. The authors find that the simulation model is sound and reliable enough to be partially valid, especially the element of the model that predicted unidentified *C. taxifolia* settlements. The authors conclude that problems do exist with the model, but that with additional simulations at other sites, the model will become more reliable.
Integrodifference Equation models and Integrodifference Matrix Population models. Neubert and Parker (2004) review Integrodifference Equation (IDE) models to show how they can predict spread rates of invasive species populations and Integrodifference Matrix Population (IMP) models to show how demographic and spatial models can be combined. The authors find that IMPs are especially useful for classifying individuals by stage rather than age, and IMPs are easy to use and analyze. The authors' primary conclusion is that IMPs are useful for managers because the models can help them understand the likely results of various management alternatives for invasive species.
Stochastic mathematical models. Mollison (1986) describes how stochastic models (models with discrete individuals) can be used to predict dispersal, establishment, spread, and persistence. He concludes that for species arrival, the shape of dispersal and distribution are important; for establishment, high reproductive rates are important; for spread population growth rate and mean dispersal distance are important; and for persistence, carrying capacity is important. Mollison recommends using stochastic models over deterministic and diffusion models for modeling control zones to prevent spread of invasive species.
Discriminant analysis.* Curnutt (2000) uses multiple, discriminant analyses to determine correlation between species distribution and climatic variables to predict plant species invasions. The model identifies areas in Australia, Africa, and the Americas as areas that may harbor South Florida invasive species. The model functions to match climate variables of a species native habitat to that of host habitat to predict invasions. The author concludes that climatic-matching can be an important part of a multi-level management strategy and recommends that future research focus on determining whether species live in similar habitats as the host region to which they are invasive.

Table E-1. Model and description (continued)

Discriminant analysis. Rejmanek and Richardson (1996) use a simple discriminant analysis to determine predictors of invasiveness. The model uses 10 life-history characteristics of cultivated pine species and classifies them as invasive characteristics or non-invasive characteristics. Results indicate three traits predict invasive species: short juvenile period, short interval between large seed crops, and small seed mass. The best predictor for herbaceous plants is their latitudinal range. The authors recommend this model as general screening tool for detection of invasive, woody seed plants.

Discriminant function and principal component analyses.* Mandrak (1989) uses discriminant function analysis (DFA) and principal components analysis (PCA) to compare the ecological characteristics of potential invasive species to recently invading species to determine potential invaders' response to climate change. The DFA uses a covariance matrix, and the PCA uses a correlation matrix. Analyses show that 27 of the 58 possible invaders (46.5%) are considered to be potential invaders of the lower and upper Great Lakes region. Eight potential invaders are thermally restricted to the Lower Great Lakes region, however, under climate change, their spread could be relatively swift. The author concludes that management practices such as stocking and rehabilitation implemented to maintain cool and cold water fisheries may be altered by rapid increases of warmwater species.

Ecological niche modeling. Peterson (2003) reviews techniques for ecological niche modeling, an approach that requires the following assumptions: (1) species distribution is limited by its ecological niche; and (2) a species can only disperse to an area with similar ecological characteristics. The results of the author's review indicate that the ecological niche or geographic element of an invasion constrains the distribution potential of a species. The author concludes the potential contribution of ecological niche modeling on the prediction of the potential range of an invasion has not been fully appreciated. The author also notes that invasive species predictions can be integrated with global change predictions.

Ecological niche modeling/Genetic algorithm for rule-set production. Peterson et al. (2003) use the genetic algorithm for rule-set production (GARP) to model ecological niches of and predict the geographic distribution of four North American invasive plants. GARP models relate ecological traits of areas where a species is located to points sampled randomly from the rest of the test area to determine decision rules that best describe those traits associated with the species' presence. Results show that ecological niche models based on the native range of species can predict invasions; thus, the authors conclude that GARP approach to modeling has more precise predictive power than other approaches.

Ecological niche model/GARP. Underwood et al. (2004) developed a model using GARP to predict non-native species' environmental niches in Yosemite Valley, considering elevation, slope, and vegetation structure. The results demonstrate the predictive potential of GARP for identifying potential invasion sites. The authors conclude that similar models can be developed for other national parks and that such models may increase the efficiency of fieldwork and monitoring while decreasing cost to managers.

Ecological niche model/GARP. Peterson and Vieglais (2001) outline a framework for developing projections of to identify the risk of invasion by species from a specific region. The authors describe the procedure for modeling ecological niches, focusing on GARP and describe several tests of GARP's accuracy. The framework depends on availability of biodiversity data. The authors identify future possibilities for using GARP, including using models to help create avoidance strategies based on what activities could result in invasions. They also note that it will be important to work with scientists and managers that have valuable biodiversity data.

Bioclimate envelope model.* Pearson and Dawson (2003) review bioclimate envelope models, discuss limitations, and propose that the model can be useful as a first approximation to understand climate impact on biodiversity. Bioclimate envelope models consider only climate variables of a species range and no other factors that can affect species' distributions. The authors state that it is not possible to accurately predict biogeographical responses to climate change, but that bioclimate models may be the best available guide for making policy decisions. The authors recommend a hierarchical modeling framework with climate as a dominant factor on a large, continental scale and biotic factors dominant at micro-scales.

Table E-1. Model and description (continued)

GIS. Haltuch and Berkman (2000) use multiple regression analysis to predict percent cover by zebra mussels. The analysis incorporates (1) bathymetry, (2) sediment type, and (3) Side Scan Sonar data. The model indicates that zebra mussels spread across soft substrates and transform soft substrates to hard substrates. Mussels on soft substrate may serve as a "positive feedback" for more mussels. The authors conclude that this GIS modeling can be used to predict spread of mussels onto soft substrates in other lakes and to examine species invasion dynamics and impacts across landscapes.
GIS. Le Maitre et al. (1996) use GIS models to show how much water could be lost per year if plant invasions are allowed to continue uncontrolled. Using this information, they develop a catchment management system. Five processes modeled separately include: (1) fire occurrence, (2) spread and establishment of alien plants, (3) growth between fire cycles, (4) rainfall to run off ratio, and (5) effects of biomass on stream flow. The model shows that, over a 100-year period, invasive plant species cover increases from 2.4% to 62.4%. The authors recommend removing invasive plant species to ensure water availability.
GIS. Wilcox et al. (2003) mapped *Phragmites* coverage over nine different years using aerial photos in Great Lakes region. The authors conduct spatial analysis of total area covered each year. They analyze abundance changes using geometric or logarithmic growth equations. GIS maps show distribution was dynamic from 1945 to 1999 and increased exponentially from 1995 to 1999. The authors conclude that the rapid expansion of *Phragmites* is due to decreased water levels in the Great Lakes, increased ambient air temperatures, and possibly natural and human-induced disturbance. The authors predict that *Phragmites* will continue to expand at a high rate into the lower Great Lakes coastal wetlands given the plant's level of invasiveness and predicted climate scenarios.
Regression analysis. Ricciardi (2003) uses regression analysis on data from various zebra mussel (*Dreissena polymorpha*) invasion sites to develop empirical models of zebra mussel impact. The analysis involves comparing an invader's impacts in the different regions and ecosystems where it has been introduced to determine if the invader's impacts are predictable in varying habitats. The author's results show that the zebra mussel impacts are predictable across various areas and habitats. The author recommends correlating models that relate invader abundance to physical environmental traits with models that relate invader impact to abundance. Linking these models to invasions will allow for predictions on which habitats will experience significant impacts from invasion, and management decisions can be made accordingly.
Logistic regression analysis.* Collingham et al. (2000) use statistical models of presence/absence of three weed species at coarse and fine scales. The authors evaluated ability of model at one scale to predict distribution at larger scale. The results show some correspondence between environmental factors at different spatial scales. The authors recommend modeling species at more than one scale. This is important for managers, because weed control happens at a fine scale, but understanding processes on a larger scale is important for long-term management. For example, analyses show that climatic variables affect species' ranges; thus, range may be affected by future environmental change.
Multiple logistic regression analysis. Goodwin et al. (1999) conducted a multiple, logistic regression analyses to determine if the invasiveness of introduced species can be predicted based on widely available biological data. The analyses were conducted on 165 pairs of plant species originating from Europe, where one species in the pair has invaded Canada and the other species has not (110 pairs were used in the multiple regression analysis and 55 pairs were used to test predictive ability of the regression models). Three biological attributes were used in the analysis. The authors found that geographic range of a species is a successful predictor of invasiveness, while the biological attributes tested are not. However, geographic range is likely correlated with biological traits. The authors conclude that predicting invasions on a species-by-species level will not adequately deal with the accidental introduction of species.
Regression and Akaike's information criteria. Marchetti et al. (2004) use logistic regression to determine the relationship between successful establishment and biological variables; multiple regression to evaluate the relationship of a measure of spread and the average abundance of an invasive species with biological variables; and Akaike's information criteria (AIC) as an unbiased estimate of the regression model fit. The results show that different characteristics favor different stages of invasion (e.g., establishment, spread). The authors find that human preference affects invasion, and they recommended stopping the transport and release of non-native fish to prevent invasions.

Table E-1. Model and description (continued)

Spatially explicit, individual-based simulation. Higgins et al. (1996) use factorial design and simple linear regression-factors ([1] adult fecundity, [2] dispersal ability, [3] time to reproductive maturity, [4] temporal frequency of post-fire recruitment opportunities, and [5] fire survival by adults) to quantify interactions between factors on spread rate of *Pinus* species. All but fire survival can significantly affect *Pinus* spread rates. Efforts need to focus on obtaining empirical data for the four relevant factors. Authors conclude that it is important that factor determination and model development are correlated to ensure that the model is designed to fit with the biological processes it is modeling.

Taxonomic model. Lockwood (1999) uses bionomial probabilities to predict establishment success of avian invaders. Specifically, the author creates a random estimation of the number of successful species per family using bionomical distribution. The results indicate (1) that some taxa are more likely to successfully establish themselves than others and (2) that human action (e.g., importing certain species) obscures trait-based taxonomic patterns in successful establishment. The author concludes that although the likelihood that transport can increase the amount of introduced species, this does not affect the amount of species that actually become established.

Quantitative taxonomic model. Kolar and Lodge (2001) reviewed publications that use quantitative methods to assess characteristics of introductions and of species that invade to document taxa-specific trends. The authors review 16 publications that look at release event characteristics or species characteristics for plants (8) and animals (8). Using the literature, they evaluate characteristics of introductions associated with establishment and dispersal of nonnative species and describe characteristics similar to species that invade. The results indicate that propagule pressure is positively related to establishment success; and the region a species originated is significantly associated with establishment success. The authors recommend that predictive models be broadened to include earlier stages of invasion.

Quantitative taxonomic models. Kolar and Lodge (2002) predict establishment, spread, and impact of invasive species in the Great Lakes, using quantitative, predictive risk assessment. Specifically, the authors employ discriminant analyses and categorical and regression tree analyses to identify traits of successful and failed invaders, considering stages of invasion separately. These two models also are used to assess the risk to the Great Lakes from species introduced by various vectors. Results show that quantitative models and taxon, ecosystem, and invasion stage specific data can be used for risk assessments and for guiding policy, education, and management efforts to prevent future invasions. The authors recommend that the approach they use to assess risk be applied to various aquatic and terrestrial plant and animal species.

Probability. Huston (2004) uses a dynamic equilibrium model of species diversity to address (1) the probability of an invader's successful establishment and (2) the probability the invader will become dominant in the invaded ecosystem. Productivity, disturbance, and environmental factors can be used to predict invasions. Areas with minimal productivity are easily invaded. Productive, undisturbed, and very unproductive areas are seldom invaded; the easiest areas to invade, establish, and impact are disturbed, productive areas. The author recommends understanding how the deviations in the landscape affect the probability and impact of invasions so that more efficient and effective monitoring and control programs can be developed.

Demographic model. Bartell and Nair (2004) created a probabilistic risk assessment framework to evaluate the possibility of solid wood packing material (SWPM) pest establishment. The approach addresses (1) pest life history traits, (2) suitable host availability and environmental factors that affect establishment, (3) population dynamics, and (4) implications of uncertainty on estimates of risk and risk reduction. The results indicate that small increases in effectiveness of treatment of SWPM can have significant impact on reducing the risk of pest establishment. The authors conclude that the methodology is a sound, transparent, and repeatable approach, and it can be used for assessing the probability of establishment by invasive species.

"Tens" rule. Williamson and Fitter (1996) use the "tens" rule and exceptions to the rule to answer the question "how might we explain [invaders]?" The "tens rule" states that 1 in 10 imported species appears in the wild, 1 in 10 introduced species becomes established, and 1 in 10 established species becomes a pest. Three sets of factors are important for the tens rule and deviations of it: (1) propagule pressure, (2) factors allowing species to survive, and (3) factors determining local abundance. The authors conclude the "tens rule" does describe much of what is known and that a more exact algorithm could be developed. They recommend additional studies on early phases of establishment, causes of death and how density affects these causes, and the distribution of species during dispersal.

Table E-1. Model and description (continued)

Screening system. Daehler and Carino (2000) tested three systems for their ability to screen for invasive species in Hawaii: (1) The North American "decision tree," (2) The South African "linear series of five modules," and (3) The Australian "49 questions." The North American and Australian screening systems had the best results for predicting invasive species in Hawaii, and both need only minor changes to be used in new areas. The authors recommend modifying the Australian screening systems for international use.
Multiple competing models. Adams et al. (2003) use multiple competing models to test the association of invasive bullfrogs with non-native fish. Specifically, multiple competing models are used with field survey data to rank the most important factors on bull frog distribution and abundance. Data from field experiments were modeled using logistic regression. The authors found that non-native fish facilitate survival of bullfrog tadpoles. The authors recommend that users regard the fish as a "keystone invader" in ponds or lakes that were fishless.
Neutral landscape models. With (2004) uses a neutral landscape model (NLM) to determine how landscape structures impact invasive species dispersal. NLMs are based on percolation theory, which examines flows through heterogeneous materials. The author finds that poor dispersers may spread more readily when the disturbance area is large or concentrated in space. Good dispersers may spread better with small and localized disturbance. The author recommends developing land management actions to control invasive species based on whether dispersal or demography affects spread more.
Economic model. Perrings (2002) creates a model of biological invasions based on fixed parameters (invasion rate, restoration rate) and a variable control rate. The model demonstrates that the higher the control rate, the lower the proportion of space occupied by the invasive species. In cases where the system is not controllable or observable, the author recommends control choices that reflect the precautionary approach.
Review of approaches to assessing invasive plants. Rejmanek (2000) provides a review of approaches: (1) stochastic; (2) empirical taxon-specific; (3) biological characterization; (4) habitat compatibility; and (5) experimental, which can be used to address three objectives: (1) prevention/exclusion of invasive species; (2) early detection and rapid response; and (3) control/containment/eradication. Descriptions of each approach, examples of its use, and how it can be used with the other approaches are included. The author concludes that the most robust predictions will be made using more than one approach at the same time, and he recommends that closer attention be given to habitat-specific predictions.

*Model considers climate variables and/or climate change factors.

REFERENCES

Adams, MJ; Pearl, CA; Bury, RB. (2003) Indirect facilitation of an anuran invasion by non-native fishes. Ecol Lett 6(4):343–351.

Bartell, SM; Nair, SK. (2004) Establishment risks for invasive species. Risk Anal 24(4):833–845.

Buchan, LAJ; Padilla, DK. (1999) Estimating the probability of long distance overland dispersal of invading aquatic species. Ecol Appl 9(1):254–265.

Collingham, YC; Wadsworth, RA; Huntley, B; et al. (2000) Predicting the spatial distribution of non-indigenous riparian weeds: issues of spatial scale and extent. J Appl Ecol 37(1):13–27.

Curnutt, JL. (2000) Host-area specific climatic-matching: similarity breeds exotics. Biol Conserv 94:341–351.

Daehler, CC; Carino, DA. (2000) Predicting invasive plants: prospects for a general screening system based on current regional models. Biol Invasions 2(2):93–102.

Goodwin, BJ; McAllister, AJ; Fahrig, L. (1999) Predicting invasiveness of plant species based on biological information. Conserv Biol 13(2):422–426.

Grosholz, ED. (1996) Contrasting rates of spread for introduced species in terrestrial and marine systems. Ecology 77(6):1680–1686.

Haltuch, MA; Berkman, PA. (2000) Geographic information system (GIS) analysis of ecosystem invasion: exotic mussels in Lake Erie. Limnol Oceanogr 45(8):1778–1787.

Higgins, SI; Richardson, DM; Cowling, RM. (1996) Modeling invasive plant spread: the role of plant-environment interactions and model structure. Ecology 77(7):2043–2054.

Hill, D; Coquillard, P; de Vaugelas, J; et al. (1998) An algorithmic model for invasive species: application to Caulerpa taxifolia (Vahl) C. Agardh development in the North-Western Mediterranean Sea. Ecol Model 109(3):251–265.

Huston, MA. (2004) Management strategies for plant invasions: manipulating productivity, disturbance, and competition. Divers Distrib 10(3):167–178.

Kolar, CS; Lodge, DM. (2001) Progress in invasion biology: predicting invaders. Trends Ecol Evol 16(4):199–204.

Kolar, CS; Lodge, DM. (2002) Ecological predictions and risk assessments for alien fishes in North America. Science 298(5596):1233–1236.

Le Maitre, DC; Van Wilgen, BW; Chapman, RA; et al. (1996) Invasive plants and water resources in Western Cape Province, South Africa: modelling the consequences of a lack of management. J Appl Ecol 33:161–172.

Lockwood, JL. (1999) Using taxonomy to predict success among introduced avifauna: relative importance of transport and establishment. Conserv Biol 13(3):560–567.

Lonsdale, WM. (1993) Rates of spread of an invading species – *Mimosa pigra* in northern Australia. J Ecol 81(3):513–521.

Mandrak, NE (1989) Potential invasion of the Great Lakes by fish species associated with climatic warming. J Great Lakes Res 15(2):306–316.

Marchetti, MP; Moyle, PB; Levine, R. (2004) Invasive species profiling? Exploring the characteristics of non-native fishes across invasion stages in California. Freshwat Biol 49(5):646–661.

Mollison, D. (1986) Modelling biological invasions: chance, explanation, prediction. Philos Trans R Soc Lond B 314:675–692.

Neubert, MG; Parker, IM. (2004) Projecting rates of spread for invasive species. Risk Anal 24(4):817–831.

Pearson, RG; Dawson, TP. (2003) Predicting the impacts of climate change on the distribution of species: are bioclimate envelope models useful. Global Ecol Biogeogr 12(5):361–371.

Perrings, C. (2002) Biological invasions in aquatic systems: the economic problem. Bull Mar Sci 70(2):541–552.

Peterson, AT. (2003) Predicting the geography of species' invasions via ecological niche modeling. Q Rev Biol 78(4):419–433.

Peterson, AT; Vieglais, DA. (2001) Predicting species invasions using ecological niche modeling: new approaches from bioinformatics attack a pressing problem. Bioscience 51(5):363–371.

Peterson, AT; Papes, M; Kluza, DA. (2003) Predicting the potential invasive distributions of four alien plant species in North America. Weed Res 51:863–868.

Rejmanek, M. (2000) Invasive plants: approaches and predictions. Austral Ecology 25(5):497–506.

Rejmanek, M; Richardson, DM. (1996) What attributes make some plant species more invasive? Ecology 77(6):1655–1661.

Ricciardi, A. (2003) Predicting the impacts of an introduced species from its invasion history: an empirical approach applied to zebra mussel invasions. Freshwat Biol 48:972–981.

Ricciardi, A; Rasmussen, JB. (1998) Predicting the identity and impact of future biological invaders: a priority for aquatic resource management. Can J Fish Aquat Sci 55:1759–1765.

Suarez, AV; Holway, DA; Case, TJ. (2001) Patterns of spread in biological invasions dominated by long distance jump dispersal: insights from Argentine ants. Proc Natl Acad Sci USA 98(3):1095–1100.

Underwood, EC; Kilinger, R; Moore, P. (2004) Predicting patterns of non-native plant invasions in Yosemite National Park, California, USA. Divers Distrib 10(5–6):447–459.

Wilcox KL; Petrie SA; Maynard LA; et al. (2003) Historical distribution and abundance of *Phragmites australis* at Long Point, Lake Erie, Ontario. J Great Lakes Res 29(4):664–680.

Williamson, M; Fitter, A. (1996) The varying success of invaders. Ecology 77(6):1661–1666.

With, KA. (2004) Assessing the risk of invasive spread in fragmented landscapes. Risk Anal 24(4):803–815.

Vanderploeg, HA; Nalepa, TF; Jude,DJ; et al. (2002) Dispersal and emerging ecological impacts of Ponto-Caspian species in the Laurentian Great Lakes. Can J Fish Aquat Sci 59:1209–1228.

Vermeij, GJ. (1996) An agenda for invasion biology. Biol Conserv 78:3–9.

www.ingramcontent.com/pod-product-compliance
Lightning Source LLC
Chambersburg PA
CBHW080634180526
45168CB00008B/3162